I Love My Computer Because My Friends Live in It

stories from an online life

JESS KIMBALL LESLIE

RUNNING PRESS
PHILADELPHIA

Books published by Running Press are available at special discounts for bulk
purchases in the United States by corporations, institutions, and other organizations.
For more information, please contact the Special Markets Department at
Perseus Books, 2300 Chestnut Street, Suite 200, Philadelphia, PA 19103, or call
(800) 810–4145, ext. 5000, or e-mail special.markets@perseusbooks.com.

Print ISBN: 978-0-7624-6171-4
E-book ISBN: 978-0-7624-6172-1

Library of Congress Control Number: 2016TK

10 9 8 7 6 5 4 3 2 1
Digit on the right indicates the number of this printing

Cover design by TK
Interior design by Ashley Todd
Edited by Jessica Fromm
Typography: Futura and Times New Roman

Names and identifying details of some of the people portrayed in this book have
been changed.

Running Press Book Publishers
2300 Chestnut Street
Philadelphia, PA 19103–4371

Visit us on the web!
www.runningpress.com

For Beckett and Lauren
For Dad

the manager. Oh yes, Aaron! What a talent. Very busy. So hardworking. You want to speak with him? Let me go get him. Oh, wait a minute, I'm so sorry, but Aaron can't come to the phone right now—he has three ten tops. He has buffet detail. He's restocking the make-your-own-salad station, and we are completely out of ham cubes. *Aaron, we're out of ham cubes!!!!!!* Anyway, what on earth would I do without Aaron? He is simply a star here at the Ground Round.'"

Unlike my friend Aaron, who used the largely computer-free 1990s to his own advantage, I was not built for the decade. I was nothing but a closeted lesbian with a bowl cut, constantly faking sick so I could stay home from school and email with my only friends, an assortment of opinionated, prolific shut-ins scattered across America and several decades my senior. Were I a 2010s kid, I'd probably have a really strange Tumblr and a small smattering of followers spread across the colder and therefore far odder parts of the globe, but at least there would presumably be people my own age who'd *get* me. Because I was growing up in the 1990s, however, it was just me and my sale-rack Abercrombie & Fitch track pants, counting the hours until school ended and I could run home to be greeted by my precious spiritual rallying cry of "You've got mail!"

Over the past twenty years we've used the Internet to connect with strangers and friends in a hundred different ways: we've ranked our "Top Eight" on Myspace, been "matched" with the potential loves of our life on Match.com, courted "followers" on Twitter, and gotten to know details about the lives of our high school classmates via Facebook and Instagram that prior generations could never have imagined knowing. It's all happened so quickly that we rarely even talk about what all these services and stages of communication meant to us as flawed, baffled human beings who, like Katie Couric, didn't always appreciate the magnitude of what was happening right before our eyes.

It's time to take a moment and reflect on the Internet that came before us. Consider this book (which I presume you are reading on a toilet) a moment of silence, a chance to go back and remember all the tireless modems connected to phone lines, working at the continual peril of someone else picking up to make a call. Rest your smartphone in the trash and conjure the feeling of your former Compaq, Gateway, or Dell computer's inelegant contours beneath your hand. Recall the intimidating weight of its accompanying monitor, which would surely have killed you if it fell on your head. Remember the impossible spatial requirements of the Trinitron monitor on your desk and how you couldn't fit anything else on that desk, not even if you really tried. Think back to when keyboards made loud, concentration-destroying noises and printer ribbons screeched until their final screech, which announced they'd run out of ink and paper halfway through producing your very important book report. Close your eyes and see that bright yellow, gay-friendly *ENERGY* logo and accompanying gay star that computers always showed when they were booting up. What the hell did that mean anyway?

The BIOS book screen.

It's been a great ride, starting with those earliest days of America gathering around its locum campfire, meeting weirdos in chat rooms, stalking crushes via Buddy Lists, and checking again and again to see if a particular AOL email was read or unread. Now that we've fully traded attachments for Dropbox, the Hilton for Airbnb, and driving for Lyft, it's time to think of what the world would be like if the Internet really had been some hobbyists' fad—if *Newsweek* had been right after all.

1980s and 1990s TECHNOLOGY, REMEMBERED

The 1980s

Eastbury Elementary School, Glastonbury, Connecticut, circa 1988

This is my earliest technology-related memory: my dad decided it would be fun to hook two computers up so that people could "talk to each other" over a tiny network. User 1 could type a message to User 2, and vice versa. This was a stunt that required using private bulletin board software, and one computer had to call the other with a modem. The majority of the students were underwhelmed by my dad's experiment. "Like this'll take off!" they concluded with a shrug.

By 1989 my brother, sister, and I were fierce loyalists to the Oregon Trail, which taught us the invaluable lesson that no matter how much buffalo meat a family ate, it could never make them sick.

Playing the Oregon Trail was an official, teacher-sanctioned learning activity in the late 1980s. There were planned periods in our school days when the teachers would clap their hands and say "Okay kids, time to go to the library and play the Oregon Trail!" and off we would

The Oregon Trail.

shuffle to spend time with a game that pandered to America's most damning, over-the-top stereotypes. Rich bankers were more special than poor farmers, the game said, and Native Americans existed only to guide your wagon down the river in exchange for $5. Bullets? Those weren't dangerous—those were your salvation. Back in the Reagan era this was what learning was all about.

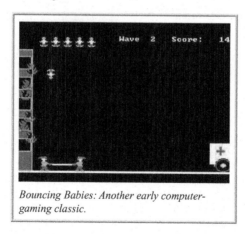

Bouncing Babies: Another early computer-gaming classic.

The other game that was absolutely revered within our home was called Bouncing Babies. In this game it was the player's job to move a safety net underneath the babies falling head-first from a burning building, thus "bouncing" them to safety.

Our computer time wasn't limited to playing games, though: we sometimes used Microsoft Word to complete our homework assignments, attempting to "save" our progress on floppy disks. The "saving" part didn't often work out. In the 1980s your relationship with the intellectual property you created on a computer was best described as a very casual fling. Nudge a computer ever-so-slightly while it was trying to save a file onto a disk, and your work would disappear.

The 1990s

By freshman year devices started getting portable. A couple of my classmates began carrying glamorous Motorola phones, which well-meaning helicopter parents had purchased under the guise of "personal safety." Such phones were used exclusively to contact boyfriends.

Other peers had use of their dads' old ten-pounder suitcase phones, which

Pre-smartphone devices. The Palm Pilot on the right was mine. The Compaq on the left, capable of sending and receiving data, was my father's most beloved accessory.

they did not ever use in public, not even in a car accident.

For once in my life I was one of the cool kids: I not only had a Palm Pilot, I carried a non-suitcase phone. My first cellular phone was a of whom didn't have official names or personalities and therefore were converted into learning tools more easily than other family pets. Come spring thaw my mother would grab a shovel and pace about the yard, trying to remember where all of her graves were. My sister and I kept hamsters, but we made certain to name them incredibly cute things like Lady Cuddlefluff and Aunt Steve as a sort of protective shield from my mother.

Both my siblings were early entrants in the school's Gifted and Talented Program, excelling at school and making themselves quite busy from grade two on. I spent my free time following the career of Bette Midler with all the restraint of an accomplished serial killer. Sure, I had other celebrity interests—mostly Midler peers, female performers from the 1970s—but Bette, to me, was the reason the world woke up in the morning.

Even though I possessed the interest set of a fifty-five-year-old, I still had the sway of an eldest sibling, so when I started talking about a secret project, my brother and sister wanted in on the plan. Initially I didn't explain the project's motive, which was to construct a world in which I was popular and my interests were lauded; I simply said that the three of us were going to build a town out of Legos and then play with it together all summer. Long accustomed to being completely ignored by me, both my brother and sister cheerfully signed on and set to work, following my instructions.

Lego Town took an entire month to construct, and when we were finished, the village traversed an entire coffee table. I taught my siblings that in Lego Town, just like in Disney World, if you could dream it, you could do it. Want a gas station in the center of Main Street next to the tree park? Grab some bricks, stack them into a hollow cube, and make it happen. A space station? We had two. A medieval parallel universe for weekends and birthday parties? Of course. A river with fish and a dock? That's impossible—everyone knows that rivers always look stupid in Legos, even if you use all the clear bricks.

The deal with playing in our Lego Town was that I pretended to be the various celebrities I admired, most frequently Bette Midler, and my siblings portrayed the normal American people whose lives were blessed by proximity to the famous. Imagine the thrill of a Lego citizen who stopped by the town police station only to bump into Bette Midler, Carly Simon, or Stevie Nicks! Although my siblings didn't care, they could not bring themselves to leave the ether of an elder sister's attention. For weeks they dutifully feigned interest in my interests. If not particularly visionary with her character development, Annie was a reliable and hard-working citizen of Lego Town who sycophantically ingratiated herself with my plastic elite. In return for her loyalty I eventually permitted her to prosper, looking over her own grocery store, gas station, and multiple-room family home on Main Street. My little brother Rob's Lego family lived in abject poverty on the outskirts of Main Street at the very edge of the plywood coffee table next to a small but permanent stain made from dog vomit. Rob's family was called the Kabobs. Rob's Kabobs. They were a collection of unemployed vagrants who existed mainly for the purpose of being beat up. On their best days they filled out the crowds at Bette Midler performances.

As the summer wore on, my siblings tired of living among my Lego celebrities, especially Lego Bette Midler, who almost never paid her bills at my sister's grocery store and once refused to visit the gas station for weeks while claiming that her motorcycle "didn't need gas." Lego Stevie Nicks didn't know when to stop performing, Lego Carrie Fisher frequently required hospitalization for "exhaustion" (I was often inspired by source material). The day things really came to a halt was when I announced that Lego Town would soon be closing all businesses for the afternoon in order to celebrate Lego Bette Midler's birthday.

upon nothing but one-sided, hyperliterate praise during such periods. I saw myself as the unofficial policeman of the group, making certain all contributors had only positive things to say about Bette. When things got out of line I sent quiet, respectfully worded email interventions to the individuals whose messages were in question, and their posts were quickly removed. In exchange for these noble gifts to my community I gained not only friends but also social status. "Jess is the only one around here with any integrity!!! It's so easy for us to forget what being on the Bette Midler Fan Message Board is all about!!!" wrote a board member who identified as SpinninDelores41, no gender specified. I didn't rudely ask for any other specifics about my friends' lives (age, occupation, length of incarceration, etc.), but after two hundred emails or so I was beginning to suspect I had a pen pal who was actually born in 1941. That meant I was a 1990s middle schooler with a buddy who could almost remember World War II.

Eventually, somewhere around year four, we tired of talking only about Bette Midler. When every era of her life had been well traversed in Friday-night private chat room sessions, we began turning inward and talking about ourselves. In emails my Midler friends started opening up about their own problems, and boy, were they weighty for a fourteen-year-old. This was before you could just Google your way out of ignorance's cave. If I wanted to write a sympathetic email to an acquaintance with a problem and thirty years on me, I had to get my information from the only Google I had: my parents.

"Mom, what happens if you miss your fifth car payment?" I'd ask casually as she prepared an all-organic dinner.

"Excuse me?"

"You know, if your"—I was now struggling to look at the notes I'd written on my hand—"if your Western Union didn't come and you miss your fifth car payment. That."

I can't even explain how badly a question like this went over in the Kimball house. Kimball rules dictated that you bought an eight-to ten-year-old Volvo at exactly its Blue Book value and then drove it into the ground until it had Flintstones wheels and a bird's nest in the engine. There was no car payment of any sort to be discussed.

"Where did this question come from?" she asked.

"I'm doing research for a paper on car payments," I'd try. "For economics."

"You don't take economics."

"It's the economics section of my drama class."

I don't know what I was actually like to talk to as a twelve-to eighteen-year-old, but I thought I was doing a pretty good job of earning and keeping the trust of my much more grownup online friends. I didn't outwardly lie to them about my own situation; I just didn't usually mention any of the things in my life that weren't things we all had in common, like scientist parents, or piano lessons, or school, or . . . anything. I was endlessly interested and sympathetic; I was always available. Somewhere in the back of my mind I recognized that I was making some sacrifices, but these were the first friends I had who didn't have yellow detachable Lego heads. This was what an online message board was before the year 2000: a place for people who needed to create their own private universes but didn't have the intellectual fortitude to play Dungeons and Dragons. Living in the online world meant I had nothing to do with the real one. I didn't care about school dances or couples or vacation plans. When I was at school I was wondering whether, upon sign-on that afternoon, my inbox would be sad, blue, and silent or bright, yellow, and affirming with a clearly enunciated "You've got mail!" by a nonthreatening man.

Eventually I decided to do the reasonable thing: I saved up all my money, bought a plane ticket, and headed out to Oklahoma to meet up with one of my absolute favorite Internet friends, BathhouseBette4928. I carefully planned my outfit (Abercrombie & Fitch T-shirt reading "I Could Be Your Worst Mistake," cropped enough to reveal my soft belly; Abercrombie bellbottoms; and gigantic, faux-wood sandals) and took off on my first flight alone. I had my parents' hesitant permission. I was weeks away from college, and I remember feeling really cool. I also remember the distinct sensation that someone should be stopping me. But I found my way to the mall where we were going to hang out, go shopping, eat lunch, and talk about Bette, and then I got stood up. I still don't know what happened; the two of us never really emailed again after that day. I heard some rumors, but mainly I went back to collecting rare photographs of Bette, trading bootlegs, and reading up on other celebrities of the 1970s. I became more cautious. I started asking people what their real-world names were, first and last.

I'd been stood up because BathhouseBette4928 and I were too different offline; I'd probably been stood up for my own protection. What an odd pair we'd have looked like, wandering around a mall with our vast age difference, the kind that would make any cop do a double take. Though I hated to admit it, I was kind of relieved. At the time some of us felt the need to meet each other in real life because it seemed like the next logical thing to do, like proposing after years of dating, but the truth was that a lot of times it wasn't.

Toward the very end of my senior year in high school something truly unexpected and strange happened: I met an actual, normal person on the Bette Midler message boards, someone who wasn't trying to mask his identity in a cloak of question marks, hide an epic weight problem, or shrug off a known status as a tax evader. His name was Marc Shaiman, and he was a musician and comedy writer. Marc was a Bette fan too, but he'd actually collaborated with Bette for years.

He'd stumbled upon our world of Bette fandom and managed to find it sweet, if extremely overzealous.

I can't remember whether it was me or my dad who eventually had the idea to just up and email Marc Shaiman, write him a note, and tell him how much I enjoyed his comedy writing and his music, his whole career. (It was probably my dad's idea.) A few hours after I hit "send" a funny thing happened: Marc responded. Believe me, it wasn't my writing skills that won his attention—I'd sent the kind of too-combed-over letter that would have benefited from a few thousand less rewrites; instead, Marc's response was a testimony to his own kindness. It was also a gesture that would change the course of my whole life.

While I remained a high school student who spent way too much time on AOL, Marc sent me a sum total of fifteen incredibly age-appropriate emails. Once, while writing music and jokes for *South Park: Bigger, Longer and Uncut*, he would email me "What rhymes with uncle!" Of course I had no idea. When, several months later, the word "uncle" made its auspicious appearance inside one of the film's very best songs, I would practically collapse in my movie seat. *I knew that word was going to be in there!* I marveled to myself, flush-faced, experiencing more emotions than an investment banker with a solid inside tip. I was, by my own estimate, now an indelible part of the constellation of great Hollywood moviemakers.

Through our correspondence Marc assured me that cool people liked Bette Midler. He also told me that there was a place just like the Internet out there in the real world where people could be interested in very specific things and even thrive because of said interests. This place was called New York City. I was accepted to NYU later that year.

Once I arrived in Manhattan, Marc and his partner, Scott, took me in in the most unbelievable sort of way: they simply treated me like I was their family. And theirs was a fabulous family to be included in:

they flew me first class to Los Angeles, brought me along to awards shows and their after-parties, and invited me to every party. (Some people joked that one could hop into a New York City cab and say, "Marc and Scott's, please!" and the driver would know where to take you.) One summer they basically gave me their beach house to use by myself for a month, plus their BMW. Because of the Internet, I went from being a nerd during my teenage years to having the social life of an heiress during my twenties. I did not wear my fortune well.

A few months after graduating from college I ended up working for a big celebrity, a very close friend of Marc and Scott's. By then the last thing I ever wanted to read was an *US Weekly* and worship at the altar of the total stranger; all my energy for consuming such details had long ago been largely exhausted. My employer was an absurdly kind person who lavished attention

Actual photo.

and encouragement upon me; one night she even invited me to tag along with her to Marc and Scott's house for a Golden Globes viewing party. With the exception of myself, the attendees were all comedy legends from various decades, people who were often either onstage at such awards ceremonies as winners or honorees or backstage writing for them. Every person there would qualify as Very, Very Famous, and there was nothing passive about their presence in a room. You invited them to a party and got a performance.

"Oh man, I hope I can keep up with this group," my boss whispered to me after I'd spaced out for a while, perhaps dazed by the sight of famous people chewing (*Stars—They're Just Like Us!*). She was huddled next to me on the couch. I was on her left. We were sharing a plate of cheese and crackers. My boss gestured to her right. "I'm seated next to Bette Midler!"

I almost passed out on the couch. Marc had seen the whole moment coming, of course, as he was well aware of how I'd felt about Bette for my whole life. Marc was smart enough not to call my attention to her presence at the party beforehand—God only knows what I would have selected to wear. In my mind's eye I can see myself and my Capital One credit card racing out to buy some tragic new version of myself at the Bloomingdale's sale rack as I prepared some awful thing to blather to her. *My Internet friends and I once sent a five-foot-by-five-foot box of fan letters to your record company, but no one ever responded.* Thankfully I had nothing prepared, so I sat, frozen, hilariously comforted by the familiar presence of my world-famous boss and the physical barrier she created between me and Bette Midler.

Marc, however, liked a laugh, and he wasn't going to let me get away with surviving the evening so quietly. He often delighted in introducing me to his famous friends by announcing, "Jess is here because of the *Internet*." He would then point at me and let the information float dead in the air. "We met online," he'd continue, "and now she's here. With *you*. Can you believe it?" The celebrities in the room would laugh wildly, delighted by the notion of one of their own making friends inside a computer instead of on a yacht or a movie set. I took my cue and hammed it up, making a face like, *I know, right?*, though deep down I wished my Internet friends could see me now. The world of the Internet chat room had delivered me—me, Mayor of Lego Town, unofficial Chief Bette Message Board Moderator, daughter of the mathmagician—to this glamorous apartment, two seats away from the very person for whom I'd gotten online in the first place. For the rest of that evening, as our party's attendees heckled the Golden Globes' winners and losers from afar, Bette and I became friends in the most wonderful, ethereal, New York City sense of the word. I laughed at her jokes, and when I dared to make a joke of my own, she laughed generously in return. After I had my moment in the sun Marc stood up and shouted one of his go-to instructions for young people in his Hollywood promised land: "Exit on that laugh!" And I did exactly as I was told. Marc was not, I instinctively knew, shouting at me to be dismissive or to earn a second laugh on top of my own. Oh no. Marc was gifting me directions to the life boat. Once you got lucky in a room like that one, it was time to shut the fuck up and learn something. These people didn't end up in this room because of a modem; they honed their communication skills on *Carson* and *Lettermen*, in front of America. They didn't want to politely grimace through some World Wide Web kid's deflated follow-up. The whole exchange—and probably something about Marc's Napoleonic command of the room—sent my famous boss into a fit of giggles. I am glad to say that in a rare moment of twenty-something wisdom, I knew to laugh too.

At the end of the Golden Globes party Bette and I said good-bye to each other in a polite tone that acknowledged that we might see each other again but probably wouldn't. Still, *Who knows*, I thought. After all, this was New York in the age of the Internet. Maybe, someday, we would.

The Glorious Inelegance of the 1990s Family Computer

MY wife is a PC person, and I am an Apple person. To my chagrin, Lauren keeps an old Dell desktop in our bedroom, a blight on our otherwise thoughtful decor, and any time she sits down at the thing I launch into a freeform lecture on the numerous merits of Apple. After all, is there anything that Apple *can't* do? Elegant stores made from glass? Simple white cords that get permanently dirty in four seconds? Cat entertainment? (Yes, it's true, our now-deceased family cat had its very own iPad app when it was alive: an animated koi pond that was permanently stocked with beautiful, digital fish. The good people at Friskies built the cat app, presumably as a marketing effort. Comparatively, the only time a cat has used a Microsoft product was when it was hooked up to a bunch of electrodes, an unwilling participant in some sort of Russian space experiment.)

Sometimes, though, as I lurk behind whatever Apple-branded miracle maker I'm using, I look over at Lauren's awful Dell and experience a very particular type of longing because I begin to miss an old friend of mine. I begin to miss Bill Gates.

Unlike Steve Jobs, Bill Gates has never done anything that was simple or elegant. Bill Gates is approximately as artistic as a math teacher who once went on an affordable bus tour through Europe. If there is ever a simple or elegant way to do something, Bill Gates does it the other way. Accordingly, in the early days of computing, no Microsoft product ever "just worked" when you took it out of the box. You took their shit out of the package and braced your body for a whole world of hell. Windows 95 in particular should have arrived strapped to a box of Franzia. Instead, it came plastered with exuberant, highly reflective stickers written in their own patented mumbo-jumbo language—"Windows 95, now with Telecommunications® and Drawing Tools™!"—all of which was way too excited about the company's latest home-computing advancements.

To be fair, the constantly-breaking-down situation wasn't entirely Microsoft's fault, as their software suffered the misfortune of running on Dell, Compaq, and Gateway computers. These were brittle, nervous machines that could be brought to their knees by almost any alteration in their physical environments. Despite owning a multitude of surge protectors, my family frequently lost hundred-dollar modems to Mother Lightning when I was a kid. I remember the way my father would bound up the stairs two at a time at the first thunderclap of a summer storm, racing to turn the computer off and save it from a debilitating unplanned shut-down. Sometimes he would immediately set to work trying to fix the thing as the thunder danced in the background, taunting him from afar. Once in a while Dad would get lucky and manage to right the lightning-zapped modem in minutes; other times his repair attempts could last through multiple business quarters. We knew never to ask him "how it was going" and instead used our powers of deduction to quietly chart his progress for ourselves, like the children of novelists. While he worked, we three kids took turns sitting steadfastly by him night after night, our eyes glazing over as he did the same things time and again: *route print, route print, ping 1.1.1.1.* At approximate mealtimes one of us would ferry up from the kitchen cashews and orange juice or other combinations of his favorite snacks, hoping to keep his strength up. For days or weeks we would wait, hoping he would soon bring our mind-numbing computer games and treacherously slow Internet back to us, thus pressing restart on our normal life.

During my early childhood there were always many computers in our house, but they were all what my father classified as "projects"—that is, inimitable heaps of an engineer's broken toys. My personal favorite from the early archives was a Toshiba laptop that came with its own branded metal suitcase. The monitor was a study in reds: there was a burgundy-colored screen background and a lighter, bright red color for the font. There was something deeply unsettling about only being able to view one's own writing in red, like you were Karl Marx toiling late into the night. Nevertheless, my dad loved that machine.

My father is an engineer's engineer, so when he needs something, he makes it. He collects computers for components, scrap wood for furniture making, cars for parts, and old tools for construction. On an indulgent weekend he will savor the adventure of a good estate sale or a meandering trip to the dump. When eBay came to be, it was as if my father's personal prayers had been answered: the world's largest junk-yard, brought to you by computers? Christ had welded all of my dad's interests together. The man basically hasn't set foot inside a brick-and-mortar store since 1997.

At that time I was just as brazenly unappreciative of my do-it-your-self family and my father's myriad talents as any lucky child is of her own riches. By my logic the "best" families were the ones who bought things in stores, the way all of my classmates' useless dads did. My friends' dads either sold insurance or were lawyers for guys who sold insurance. Comparatively, these chumps didn't know how to make toast, let alone rebuild a car engine. On weekends they wore the designer polo shirts their wives picked out for them at Filene's Basement, and they talked a lot about having joined the country club or wanting to join the country club. My father wore his grease-stained work clothes and running sneakers, and at dinner he talked a lot about whatever he wanted to learn next. He has always been one of those people who looks around at the planet and sees Hogwarts. Again, because children are horrible, I was embarrassed by my father's self-actualized resourcefulness and therefore lied to my classmates about the origins of our objects when they came over to visit.

"Look at that coffee table!" exclaimed some kid who'd never seen an unbranded good in his life. "It also holds pictures? That's so cool! Did your dad make that?"

"No," I said forcefully. "We bought that in a store."

"Well, I've never seen anything that fancy in a store! That's just beautiful!"

"It's there if you look."

Although my dad is a man carved in the shadow of Ralph Waldo Emerson, he's still a computer engineer by trade, and by 1995 the 1990s' fervent excitement over the ever-modernizing family computer had finally started chipping away at his minimalist morals. After all, this was his epoch: he had spent two decades believing computers would someday change the world, and it was finally happening. Television ads depicted kids happily doing their homework aided by the knowledge reserves of a CD-ROM encyclopedia. Businessmen booked plane tickets to Dallas and played countless rounds of poorly animated Blackjack just by clicking mouse buttons. Soccer moms dutifully organized their grocery lists by aisle with an Excel spread-sheet. It was a whole new universe.

"Want to come with me to CompUSA?" my dad asked me casually one morning in June, as if he had just asked me to help with yard work.

I was already an active AOL addict at the time, but I was running a very old version of the software and Windows 3.1 on one of my father's former office computers and was years behind the setups of my closest online friends. No one else had a lousy 14.4K modem or a measly ten-inch monitor or a ribbon printer. I wanted a machine that could not only download Bette Midler pictures but scan them as well. Of course I wanted to go to CompUSA!

An hour later my dad and I were standing in the hallowed parking lot, and as he moved his toolbox to make room in the trunk for a poten-tial purchase, he warned me that we would have to be on the lookout for a very good deal. If there wasn't a good deal, then maybe we wouldn't get a computer today at all. I nodded to let him know I understood the gravity of the situation. I knew my dad, and I would fundamentally disagree on which computer to get, as I would want the most glittering, pointlessly expensive machine available, and he would want whatever "made sense." We had everything and nothing in common.

Once in the store my father silently paced around the home-com-puting section for over an hour while an eager, commissioned sales kid tailed behind, attempting to decipher his emotions. When my dad is unhappy with his circumstances and trying to think his way out of them, he looks like he's trying to smell the air somewhere very far away. The computer cabana boy had no idea whether this expression was a good or a bad sign, but I knew my father well enough to identify his sentiment's exact makeup: it was part disappointment with prices and box retailer bundling strategies, and part resignation that he would most definitely be roped into paying for features he didn't even want.

To my father, it seemed that the store delighted specifically in swindling computer engineers. He was pacing because he was attempting to make peace with himself; he was running through opportunity cost calculations in his head.

Two hours later and without the sales boy's help, we left CompUSA with a gloriously huge box. My dad had settled on a Dell computer, and I couldn't believe this was happening to me. Back home my brother and sister had prepared welcome decorations for our newest, most beloved family member. A desk had been cleaned and printer paper located in the hopes that Dad was also bringing home a new printer (which he was). Upstairs, with the new computer settled in the study, my dad set to work. Back then, installing Windows 95 required loading and running dozens of hard disks in the right order: make a mistake, and you had to start completely over. It was like a Microsoft-sponsored edition of Chutes and Ladders.

Windows 95: When you had to watch an hour-long instructional video before you could successfully install a piece of software.

passwords that blocked us from them. Because of this, my brother, a nine-year-old who looked like he was on a hunger strike, could sit down at your computer and repartition your entire hard drive: "Open a prompt, you're in C:\. Enter DIR, that means directory—see, here's a huge list of files and folders. DIR /p, and you can flip through page by page. Oh right, he put it in Program Files. That was obvious. CD [change directory] Progra~1. CD id. CD doom. dir *.exe. doom.exe. I found Duke Nukem. It's right here."

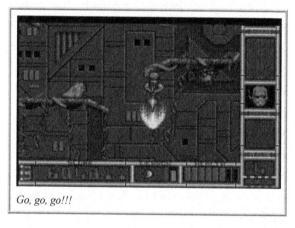

Go, go, go!!!

Today, attempting to find your computer games or fix your hardware for yourself is unheard of. All Apple products are sealed in cases that are almost impenetrable to their human owners. Back in 2007 not being able to open a backdoor to your phone's innards was a huge, huge issue. "What?!" the tech reviewers cried in unison. "We can't open the iPhone and dick it up all by ourselves?!" After all, they reasoned, it was *your* goddamn phone: if it broke, you should be able to open it up and stick fingers or glue or rice or sand or crackers or whatever the hell you wanted inside it to fix it. Nowadays my kid doesn't even know that at-home device repair was ever an option to mankind. "This is broken!" he confidently shouts as soon as our iPad runs out of juice and stops letting him cut fruit in half.

The critical difference in today's games is not in how realistic they now look but in how infrequently they break. When something works almost all the time, we become immune to its complexity and the beauty of its simplicity. We forget that computers are nothing more than brilliant combinations and permutations of zeros and ones. Modern-day computer users no longer require a working knowledge of command-prompt language just to connect to the Internet or uninstall a piece of malfunctioning software. No longer do we experience that gorgeous, raw moment of "pinging" a website—the computer code equivalent of saying "Are you out there?"—and seconds later seeing that website ping us back: "Yes, I am." The most magical things I ever saw on computer screens were things like that, like the working ping: the things that happened in gray or green type against a dark black background, the Internet at its barest and best.

My 2015-issued MacBook is great in its slick practicality, for sure. But when I look over at Lauren's monstrous PC, I'm taken back to those halcyon days of computers when I couldn't help but get kind of excited when our ugly old Gateway suddenly didn't turn on and my dad sat down to the work of wondering why, with us three kids at his side.

Yahoo!,
I Think I'm Gay

O N the very best day of my childhood my mom treated my sister and me to a designer clothes shopping spree at the Limited Too. In the 1990s the Limited Too manufactured the fluorescent-colored, faux-hippie outfits donned by suburban America's coolest kids. After years of being forced to wear whatever emotional atrocities the monsters at Land's End had sewn up, my sister, Annie, and I wanted in on the real action. At long last my mom had relented, throwing her values to the wind and driving us thirty minutes to the neighboring town's expansive mall, Buckland Hills. After many giddy, calculating laps around the store, I decided on a pair of teal jeans, a plastic choker necklace, and a coordinating teal-and-pink T-shirt featuring a colossal foam-painted Limited Too logo across the front.

"Did you pay for that?!" was my father's sputtering reaction when I arrived back at home, he a man of battered Levis and free T-shirt souvenirs from antique car shows and regional jazz music festivals. "You *paid money* for the privilege of being a national clothing company's walking billboard," he scoffed. Clearly my father thought he'd learned me better, but alas, there I was, a penniless, potbellied spokesmodel for the corporate tricksters at the Limited Too.

The next Monday morning I hoped my outfit was sending a loud and clear message to the Connecticut middle school world: I had arrived.

Sure, that arrival had entailed rapidly wheeling my string bass down the hallway and being extremely late for first period orchestra class, but mine was the kind of unassailable confidence that great fashion gifts its savviest ambassadors.

I had been desperate to look cool on the outside because deep on the inside I knew that something about me was different from my classmates. All the other girls my age were interested in boys, and I was not interested in boys. All the other girls spent their time pleading with their moms for the right to wear incrementally sluttier outfits to the next school dance, and I spent my time dreaming up excuses for why I would not be able to attend the next school dance. To me there was nothing romantic about some kid leaving sweaty paw prints on the side of my Contempo Casuals cocktail dress during the approximate sixty-seven minutes it took Led Zeppelin to complete playing "Stairway to Heaven." I didn't exactly know why I was so opposed to school dances, but, like a dog lured into a bathtub, as soon as I found myself in a packed gym blasted by Backstreet Boys music, I knew I needed to get the fuck out of there.

As a middle schooler, I didn't know that I wasn't straight because culture hadn't given me the language or the relatable example to know that I might be something other than straight, like gay, or bi, or bi-curious. The 1990s were not an era with any interest in ambiguity. As far as I understood things, you were either gay or you weren't—it was that simple. I knew that if, indeed, you wanted to "be gay," then you had to renounce yourself to a life of public otherness, like some crappy version of being a monk. Your first task would be to Come Out because everyone and their grandmother had a God-given right to understand exactly what kind of bedroom activity you were into, and then, after your huge, painful announcement, everyone in your life was perfectly allowed to weigh in, *especially* if they were upset. Because of course!

Of course what you choose to do in the privacy of your bedroom should have a profound effect on people who would never, for not even a second have to witness it. It was no wonder that I was desperate for a branded Limited Too disguise; the 1990s were only nirvana if you were opaquely, unshakably heterosexual. For the rest of us kids, these were the years when Tom Hanks was "gambling" with his good name in order to play the part of a gay man in the movie *Philadelphia*. And I quoted the *New York Times* with that descriptor, not some Bible-thumping aunt of mine.

When I began to question things, slowly realizing that my deep disinterest in prom might be about something bigger than my dislike of other kids—like, say, my disdain of my own identity—I did what I always did: I turned to the Internet.

Back in the mid-1990s Yahoo!, Lycos, Geocities, Alta Vista, and Search.com were the portals of choice for the trailblazing web surfer. Google was still years away, and anyways, the Internet didn't really feel big enough to need search algorithms yet. The URL addresses of most significant webpages were listed somewhere on a hub page. There was, already, a desire to have a place on the Internet for everything, even the gay things. Questions about sexuality were largely relegated to sites run by nonprofit organizations like Planned Parenthood or experts like Dr. Ruth. Most of these sites had the noble intention of helping young people figure out which side of that two-color rainbow they potentially sat on: homosexual or heterosexual? While you presumably figured things out, you could also buy a T-shirt with a pride flag on it, or call a national clinic to get some medication, or register to march in a summertime pride parade. Because no one knew what on earth to do with all our new, colorful, digital tools, it was also possible to download digital wallpapers of the herpes virus or turn your cursor into a sperm.

In between such practical matters as locating gay-friendly campsites and purchasing garish, rainbow-patterned clothing, these webpages created editorial content on all things homosexual. It was in these databases where I spent a significant amount of my free time. "Am I gay?" other site visitors asked the experts manning the Dr. Ruth message board. "If I'm attracted to women and women only but I'm married to a man and we have three kids, do I have to be a lesbian?" "Help, my mom found my diary and now she knows I'm gay but she's not saying anything and I'm not saying anything and we're Catholic and also she's the town mayor." God, you could feel the nervous sweat through the modem wires.

Dr. Ruth did her best to provide her specific brand of kindly, acid-tripping grandma advice to the authors of such queries, but she and her ghostwriting team could only handle so much of the stifled, anonymous gayness that was flooding the early Internet. Soon other sites curated pieces from less notable sources. Allow me to sample for you the general tone of such content through these supremely terrible headlines:

"Hope: An Interview with a Lesbian Who Lives Near New York and Is Fine, Just Fine"
"Moving On: How to Break Up with Your Homosexual Partner with Dignity"
"Mourning the Life You Would Have Had: Telling Your Parents, Friends, and Grandparents About Your Homosexuality and Helping Them Start to Understand"

And, last but not least,

"Getting Help"

It was exactly as awful as it sounds.

Soon, as the wheels of innovation began to turn, a new idea appeared on several of the Gay World Wide Websites. This particular formation of content was such a good idea that it continues to thrive today, thanks mostly to the children at BuzzFeed. I am talking, of course, about the Internet quiz.

"Could You Maybe Be Gay? A 25-Question Quiz"

Lacking the wiz-bang features of modern digital content (big data! video! QR codes!), the amateur-created HTML quiz was nonetheless a new, unprecedented way to engage with the casual web surfer wondering whether they were gay. Terrified, I waited the two to five minutes for the page to load and committed to take the quiz, no matter how scary it was.

"Have you ever kissed someone of the same gender?" NO, I selected, instantly feeling much better about where this quiz and my life were headed.

"Have you ever had a relationship with someone of the same gender?" I breathed an audible sigh of relief by the time I got to this question. NO. I was fine!!! Thank you, World Wide Web, for clearing up this horrible confusion for me. As if I were a gay person!!! I lived in Connecticut, for Christ's sake! Seven or so questions later (plus a four-minute wait for a pixelated hourglass to turn itself upside-down a few times), the quiz concluded that by its calculations, I wasn't gay. I wanted to print out the results and save them. "It's normal for everyone to question their sexuality sometimes," the quiz added.

But as I toggled my way back to the safety of America Online, I had queasy feeling in my stomach. What if I'd been too conservative in how I'd answered the questions? What if I was just *one question away* from being gay? That would be different. I decided to go back and answer the questions again, this time stretching the answers a bit to see how I fared when I was joking—totally, harmlessly joking.

This second time I checked a few more yeses on the feelings questions and confessed to more confusion and crushes. I told no lies of relationships and other hard facts. The quiz, however, changed its opinion. "You may be gay," it concluded, "but also you may not be, since most people go through 'phases' when they think they are gay."

I left the quiz website more confused than I'd started off. Thankfully, back at the ol' AOL homestead, there were plenty of chat rooms for debating the answers to such questions. The chat rooms had basic names—M4M, W4W, and, later on, M4WMWM and the lot. In these years the Internet was both global and local. It was quite possible to go into a chat room in America Online and find only twelve or so members of America Online. Those were the people with whom you then discussed the headlining topic. Here to talk about W4W? Those twelve people represented the global online totality of your options. Perhaps the weirdest part was figuring out that most people in these conversations were not confused teenagers like myself. Most people in these chat rooms were something far more troubling: confused adults.

It didn't take the losers of the early Internet era very long to figure out that in such situations, given the lack of video cams, it was best to reinvent yourself as someone cooler than your real self. Moms became teenagers, dads became moms, and teens became twenty-somethings. No one was the age they claimed to be, and the standard deviation was somewhere around plus or minus fifteen years. (For purposes of believability, it was easiest to stay somewhat within the confines of your generation, lest you be asked about how you voted in a presidential election that you had no idea even happened.) Everyone simply designed the most efficient way to explore their own fetish and then proceeded accordingly.

When your fake cool persona interacted with someone else's fake cool persona, the result was a conversation like this:

Tina3366: asl?
Frederica2: 20, W, NYC. u?
Tina3366: 26/W/TX

By now I know I'm talking to someone who likely lives near Texas, but is definitely old enough to be my parent. The "66" in the screen name is probably a better clue than the "26" in the answer. I try my best to ignore these odds.

Tina3366: Do you have a girlfriend?
Frederica2: No. Not really. You?
Tina3366: Not really.

Oh, the undiscussed worlds behind the hedged answers like "not really" in the 1990s AOL chat room. Online the women—or those who were accurately portraying women—were never overly forward, and this led to a lot of painfully casual chat.

Tina3366: What are you interested in?
Frederica2: I like music. Stevie Nicks, Pink Floyd, the Dire Straits, lots of different stuff.

Naturally, 66 percent of that answer does not represent my real favorites; instead, I am trying to create what I believe is the musical taste profile of a cool person. But because I've never spent time with a cool person, I am really grasping at straws here. Once I was a seasoned chat roomer I'd occasionally take huge risks just to see how it felt. Because I was at a computer, I assured myself, it was always possible to simply turn off the computer and abandon all attempts to socialize if things suddenly got overwhelming. Such risks usually went something like this:

Tina3366: I love all those artists. What TV shows do you watch?

Frederica2: I like Buffy the Vampire Slayer the best. I think Sarah Michelle Gellar is hot. ;)

Tina3366: Me too. :-)

Just seeing myself type something like that, about a fictional charac-ter, no less (who was, by any cultural standard, really hot), was enough to make me cringe and hold my head in my hands. What was I feeling? Did it feel more or less like me? Was anyone in my family anywhere near the computer room?

Tina3366: Have you ever been in a chat room before?

Frederica2: No, not really. I was just curious. You?

Tina3366: A few times, sure.

These tiny admissions, whenever someone admitted that maybe they were in a W4W chat room because they were curious about the "4W" portion of things, felt like mini-earthquakes. I felt emboldened and embarrassed. I had been handed a rare moment to seize: I was chatting with a person who might be real, who didn't seem to be a total nasty sociopath, with whom I could interact and try out this new version of myself in as safe an environment as had ever historically existed. This was my moment.

Frederica2: What kind of things have you talked about?

Tina3366: Oh all kinds of things come up, especially after a certain hour LOL.

Frederica2: What was your best convo?

Sometimes, often right when things started to get interesting, AOL users' friends and family would join the chat room. Such IRL interventions allowed for moments like this one:

Tina3366: Y THE FUCK ARE U TLKING TO MY MOM
Tina3366: DIE!!!!!!!!!!!!!!!!!!!!!
Tina3366 has left the room.

Apparently Tina3366's kids had entered the room when she wasn't looking. Apparently they didn't approve of me, her long-distance love. I logged out of the chat room. I didn't need this kind of baggage from her kids. I was a kid myself.

I had conversations with several Tinas, and I always took them very seriously. For the two to three days each one "lasted" I tried to walk to school imagining I was a normal person who was in a normal relationship, one more true and meaningful than those made up of the monosyllabic exchanges and hand holding I paid witness to every day in the hallways. I always imagined my paramours as the most beautiful, incredible women who'd ever used the Internet: they had bustling social lives, and it was merely a coincidence that they had entire afternoons available to spend chatting online with me about nothing. *It was so random, the way we first met!* they'd confess someday soon to their "real life" friend over a coffee, served in their immaculate and spacious Texas kitchen. *I went into a chat room, you know, just to see what it was like, and I never expected to actually meet anyone there. Like I need a chat room to make more friends, you know what I mean!?* Both women would laugh wildly. The thought of meeting up with me in person would seem silly and implausible to Imaginary Tina, but also tempting and fascinating.

When at long last I'd fly to Texas the following summer I'd have lost some weight, gained some style, and convincingly learned to come off as being five to seven years older than I actually was. Of course I'd immediately become the closest of friends with all of Tina's friends. Come evening Tina and I would share a bottle of fancy wine together and then spend the night cuddled by a fire, kissing with all our clothes on. We'd live happily ever after, unconstrained by the expectations of the economy or the needs of her children, who would miraculously disappear.

These ephemeral, largely fictionalized chat room romances were the beginning of my understanding of what it meant to be myself, and for better or worse, they couldn't have happened without the Internet. As an adult I look back and think of all the risk I was spared. Had I been just five years older, my first experiences probably would have involved some poorly lit bar a few towns away, where the police maybe monitored the parking lot and parents could be called in the middle of the night. Instead, I came of age on the very edge of the Internet frontier, and it was there that I had mortifying exchanges with Tina, or whatever her real name was, as she and I figured our real selves out.

2000-2005 TECHNOLOGY, REMEMBERED

Computers, especially their monitors, are now the size of mast-odons. Parents routinely throw their backs out while attempting to hurdle $1,100 refurbished Dell Motherfucker desktop computers onto college dorm room tables, shot-put style.

Instant Messenger, technology's single greatest achievement, enters its halcyon days. The drafting of one's away message is the 2000s' most important and personal haiku. Want to tell someone you're in love with them but also don't care if they care? Want to generally seem deep or aloof? Your away message is just the place to do it.

BlackBerrys emerge on the scene, first for rich people only and then trickling their way throughout the rest of the business world and eventually into teenagers' hands. They are an incomprehensibly huge status symbol on the New York subway. iPhones of 2007 were nothing in comparison. The iPhones of today still do not send and receive email nearly as fast as the BlackBerry did.

Brick Breaker. The greatest game in the world was Brick Breaker. Even your boss played it. Especially your boss played it.

The first startup of significance arrives in New York City: Kozmo.com. Kozmo is run by an enthusiastic and talentless businessman named Joseph Park. I admire Joseph Park because he seemed to know that he was an idiot: he ruined his company's own IPO prospects by speaking to the press during the requisite "quiet period." That is a nursery school–level business mistake.

Back in the sleepy 2000s we could never imagine companies zipping around the streets of Manhattan delivering beer, porn, dinner, and antacids to rich people until the day Kozmo.com appeared.

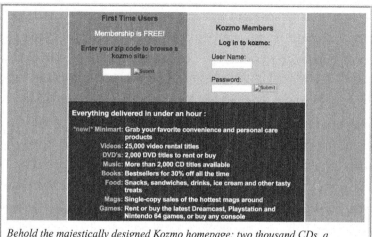

Behold the majestically designed Kozmo homepage: two thousand CDs, a sandwich, and a Dreamcast, all delivered straight to your door in under an hour.

To compare Kozmo.com to the standard technologies of the time, try to imagine a world pre-iPod. Try and remember that listening to music in the year 2000 still meant dutifully spending hours downloading illegal MP3 files from LimeWire or slowly uploading our legal Billy Joel and 98 Degrees CDs onto our personal computers, hoping to eventually toot out the songs in an order of our own choosing. We 2000s people had Winamp media players. We were absolutely stunned to see a computer's processing power harnessed for rotisserie chicken delivery.

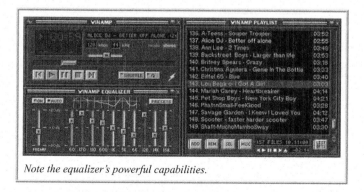

Note the equalizer's powerful capabilities.

Sensing a new zeitgeist emerging—one wherein the personal computer could become *more* than just an email-avoiding, music-uploading, and instant-messaging machine—a competitor to Kozmo.com quickly arose in the form of UrbanFetch.com, which also provided twenty-four-hour-a-day delivery e-commerce. There was perceptibly no difference between the two business models. In the great tradition of techie capitalism, the identical venture capital-backed dot-com startups faced off, setting about the work of capturing the bitchy hearts and minds of New Yorkers by doling out endless $50 coupons to new users. I think I must have redeemed at least seventeen of those little VC-backed golden tickets myself, burning through fake email address after email address as I pretended to sign up anew for Kozmo.com while the same delivery guy was dispatched to my dorm room door every time, arriving with my same dumb order of beer and pizza-flavored Combos and a polite grimace on his face. We both shared an unspoken sense that these days of glory probably wouldn't last.

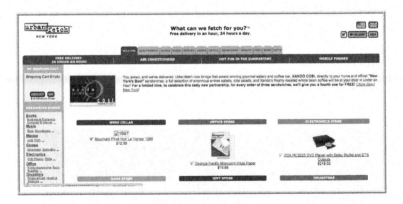

The Achilles' Heel of Myspace Tom

I N 2007 some people used Myspace, and some people used Facebook. Countless articles debated the merits and demerits of each service. "Myspace looks so unprofessionally done when compared to Facebook," mused the tech blog Mashable. "But every band ever has a Myspace account." Whoa, good point! This was going to be a challenging debate.

Music was often at the cornerstone of the "which social network is the best social network" discussion. This is because many pundits envisioned the near future of the Internet as a sort of always-open international café, a Peach Pit for the whole planet. Myspace fulfilled some portion of this fantasy because of all the musicians all over the site; their presence made the Internet significantly cooler than it was in the days of being inhabited by people like me and my shut-in friends from America Online. Facebook, comparatively, was the quiet-blue Harvard library where you might get into trouble for even bringing your music, even if you'd planned to use headphones. "What's on your mind?" Facebook gazed penetratingly into your soul and asked you as soon as you entered its pin-drop-quiet hall. And you'd better have an answer for that one.

Yes, the services were different, and yes, Myspace had webpages for bands and Facebook had news feeds for people, but the real reason

Myspace failed and Facebook succeeded had nothing to do with music. Myspace failed because it was designed by a popular person for a popular person's world, and Facebook succeeded because it was designed by an unpopular, wildly uncomfortable person for the real world.

Myspace was built by Tom, a handsome, easygoing guy who was continually looking over his shoulder at you as if he was on his way to somewhere better. "Hi Tom!" you could imagine yourself calling out across a fall football field, appreciative of the initial gesture of camaraderie that was his appearance on your Top Eight list as your Number-One Friend. "Thanks for being my Number-One Friend! I see you're busy, so, uh, maybe I'll bump into you later, dude!"

In his infamous profile photo, Myspace Tom is standing in front a whiteboard that has a blurry treble clef and a sharp note scrawled on it. There are also lyrics or song titles in confident capital letters, a set list, perhaps. *Bee bop beep skeedle-e-doo* cool musicians' stuff. That's the general vibe of the picture. It's very clear to me that Myspace Tom is hanging at a jam session. The jam session appears to be held at a school—where else would you find a whiteboard—but the teachers probably gave Myspace Tom the room for the afternoon because he was just that intergenerationally respected. "Do whatever you want in here," they probably said. "Drink, smoke, bet on horses—see if we care." Judging by Myspace Tom's cocky but endearing expression, the jam session was completely awesome. An explosive hour of covering Dave Matthews Band and Guster songs and then maybe one tune by Three Dog Night as a tribute to his parents. That's probably the moment when someone snapped the infamous profile picture of Myspace Tom, just as he called his mom on his Razor phone and shouted, "Hey Mom, remember this one?" *Jeremiah was a bullfrog!*

Myspace Tom probably had an electric guitar draped between his hands at that exact moment—he didn't use the strap—because

Myspace Tom was just that loose of a guy. Music was in his *blood*. That's because Myspace Tom probably had a cool uncle who owned a jazz club or something. He probably hung out there at a precociously young age as a kid and snuck beer. Now old enough to go to a jazz club all by himself, Myspace Tom was just hanging around with his friends and then probably going to a party at a girl's house after that. To an Internet nerd of the day, the life of a well-liked, socially adjusted guy like Myspace Tom was as unimaginable as North Korea.

When you think about the kind of problems that a dude like Myspace Tom encountered in his daily life, Myspace's hodge-podge feature set makes complete sense. *Where am I going to house all my sweet tunes, so that all my many fans can access them whenever they want to? How can I keep track of my best eight friends when I have so, so many best friends and am constantly making new best friends all the time? And they're all great? But some are greater? How can I subtly let my old best friends people know that they're not my new best friends anymore but they're still my friends-friends for sure? Also, where can I express all my ideas for the different colors, fonts, and designs that could collectively best summarize my personality in any given moment? How to I harness the myriad possibilities of me?*

Comparatively, Facebook was made by Mark Zuckerberg, a man whom I'm certain was a violin-carrying, Magic: The Gathering–playing weirdo when he was in high school. Mark Zuckerberg never had an uncle who owned a jazz club. Mark Zuckerberg was never even invited to listen to a recording from a jazz club. For people like Mark and me, the horror of having to arrange our "top eight" friends in a public list was a reason why we might lock ourselves in a closet for a weekend. Write down our *top eight friends?* What kind of popular kid-created totalitarian society were we living in now? What if we really had two friends, and one of them might not like us anymore, and one of them was someone we met at Space Camp? And we couldn't remember their names right now? What then, Tom? What the hell did we write in then?

By the end of my time on Myspace I had a total of six people in my Top Eight. I had thirty-six friends in total, but only thirty of them have added me back, a statistic made blatantly clear by the user experience of Myspace. (I didn't want to know that! And I certainly didn't want anyone else to know that! GOD, Myspace.) These popular-kid mistakes were just clearing the way for a Facebook. After all, people weren't coming to the Internet to learn the truth about how many friends they had; people were coming to the Internet for something far more palatable than that.

My Myspace Page, 2006–2007

Let's look at the kind of person that Myspace's business model left out.

Me.

My Myspace screen name is "Save Us From The Whales." Yep. I think that'll set the tone for what you're about to see pretty well.

This is probably the best I ever looked, as I am wearing more makeup than I have ever worn in my life, even at my wedding. My dress is a hideous black sack that ties behind my head in a big knot, as if the dress gave up on itself: *Well, we'd planned on the back of the dress looking nicer than this, but finance just said we're over budget by 4 percent, so, uh, this is the end.* My "jewelry," to use the very loosest definition of the term, consists of twenty plastic bracelets. I am carrying my grandmother's purse, which is full of Trident gum and dollar bills, and that is my fifty-seventh Jack and Diet Coke. I am exactly what happens when you invite an unsuccessful young person to a successful older person's party: I have cheap clothes on, I've forgotten to say thank you, and I've probably cost the place $438 in food and beverage outlay. My sincerest apologies, Scott Wittman and Marc Shaiman, and I am so proud that your Broadway and Hollywood existences have financially survived the stress of inviting me to your premieres.

The cherry on the cake of this picture—perhaps you've spotted it already and groaned—is the fact that I am standing just a few feet in front of Tom Hanks (who looks distressed by the presence of the camera, I might add, but we'll skip that detail for a minute). I *will* give myself some credit in that this picture was taken in 2007, so I'm early to the long-lasting Internet trend of finding that One Picture In Your Life Where You Are With A Celebrity and using it as your social media profile picture, hoping to cast the impression that you are *constantly* hanging out with people of such stature. *Just another day, me and Tom Hanks!* That is the message we're all hoping to send when we create a social media profile picture of That One Time When We Met Someone Famous.

The worst part of the Jess Is (Not) Hanging Out With Tom Hanks picture is that there's something about my facial expression that perfectly reveals how a person feels when they are out of their social element, the person who is the most desperate, outranked, and rightfully ignored by their surrounding company. To look at the photo is to feel both pitiful and uncomfortable, yet I have allowed myself to rewrite the narrative so substantially so as to convince myself that I look cool in this photograph that it's my Myspace Profile Picture. This photo is my silent rebuttal to Myspace Tom. *Oh yeah? You're at your band practice, Tom, because you're so socially well adjusted? Well, I am standing in line for another alcoholic beverage just a few feet ahead of my very good friend Tom Hanks.*

Also, that guy who's staring so admiringly at me, who might accidentally give the impression that I am one of the gathering's most sought-after guests? That's my cousin Greg. He's, like, seventeen years old right there. He is terrified. That facial expression is an accident.

In my next Myspace picture I am prying a bone from a dog's mouth by using only my teeth. I am wearing Uggggggggh boots, men's boxers, and a small clip in the back of my hair. The clip is the perfect way to complete to the look. The dog has a nice new collar and a haircut and looks bored. The dog also looks richer than me, which is accurate. How does a dog with only one visible accessory communicate more wealth and status than a Homo sapien? I don't know, but she's doing it.

This picture really amplifies the rest of my Myspace profile, proving that I am cool in a multitude of settings. *Oh, she's not only clearly very good friends with Tom Hanks, but she has a busy career as a dog babysitter too. Next stop, Congress.*

That dog's name was Walli Wittman. One afternoon my best friend, Lucian, and I went to Petco and decided it would be fun to make Walli a new dog collar with a new, more exciting name. We rechristened her Walli Shelby Louis Vuitton Wittman. We spent all day and $45 of our "emergency expenses" money to make with a machine a personalized collar and name tag. We put a picture of the final effort up on Myspace when we were done. That's the kind of thing that happened if you hired

me to dog babysit and you had friended me on Myspace: you went online one afternoon and discovered that your dog had a new name.

In another one of my other photos I am fake smoking a drugstore cigar and drinking tap water from a doll's tea set. My clothes don't match, my hair is stupid, and I generally look like a disgraced 1970s tennis star who has recently put on weight. I am talking with my writer friend John, who is placing a call on a fake plastic cell phone. Obviously this was another good choice for the permanent visual archive of my life to leave for my stepson to someday find of me on the Internet. So thanks for that one, Myspace.

After viewing the desperation that was my Myspace page, it's easy to reconstruct why the world was so relieved when Facebook arrived on the scene. *Welcome to Facebook. We know you don't have eight real friends, so we won't talk about that point ever, ever again. Over here at Facebook you can pretend to make friends with the thousands of people who dislike you. And because you don't even have enough enemies to round that list out, you can just add these brands instead. Brands are your new friends!*

Why try to stay friends with eight real people on Myspace when you could pal around with the Hamburger Helper on Facebook? That's how grownups understood Facebook once it opened its gates to the adults of the world. Myspace freaked us out. We needed a safe place in which to recover for a decade or two after that. And that place was Facebook.

Myspace had believed that every human being was special, and Facebook knew that we are not. *Don't worry,* Facebook said, *you don't have to be special here. You don't even have to say what's on your mind anymore. We took that prompter away.*

And thank god for that, Facebook. The world's nerds salute you and your awkward social-ish media service.

Jesus Had
a BlackBerry

WHEN I was twenty-four I worked at the sort of breezy lifestyle magazine that you find in the lounge areas of New York City's most exclusive hotels and medically unnecessary doctors. It was a glossy, guileless publication that featured articles on topics like where to shop in Paris or where to dine in Rome, as if these were fearless new avenues of discussion. Said articles were always written by someone too young and too poor to have ever been to Paris or Rome, and the stock photography competed to be as unmemorable as the writing. In general, our magazine worked best when the layout encouraged you to just keep on flipping through without stopping, sort of like when you're eating a Dorito and you don't even taste it because the only thing you want is another Dorito.

Although I occasionally contributed 250 words about David Yurman's new line of holiday-themed collectible glass dolls or 45 words about the most noteworthy spring lipsticks of Ibiza, my job wasn't on the editorial side of the magazine; my job was on the publishing side. *Publishing* is the fancy word that magazine people use for *sales.* Both the *editorial* people who write the articles and the *publishing* people who sell the ads next to them are always described as "symbiotic" and "equally important," but we all know they are not.

Richard, our magazine's publisher and CEO, was an Englishman who owned a group of American magazines that focused on European business travel—as if Americans appreciated anything about Europe. Richard was a man whose life required no fewer than four assistants. He had a main assistant, a personal assistant, a travel assistant, and a meetings assistant (yours truly), the last by far the crappiest of the crappy assistant jobs. My primary responsibility was to book, confirm, and then reconfirm the ten to twelve sales meetings with advertisers that Richard insisted be packed onto his calendar on a daily basis. Any fewer meetings in a day, and he would start to rant about his time not being utilized properly, like he was some sort of deity visiting from another universe. In reality, he was an intimidatingly tall but dough-faced magazine salesman in a shiny Brioni suit, perennially toting a portable DVD player and a ten-minute corporate video flaunting the many G-list celebrities who regularly attended our magazine's parties, including but not limited to That Chick From *The Sopranos* and "Chef" Paula Dean. My job also involved making sure that Richard's clueless, wears-sunglasses-inside personal driver knew the best driving route to take for each of the day's locations as well as which side of the street he could wait on without incurring a parking ticket. Mind you, we are talking about the year 2005, practically a decade before Google Maps would be able to answer such questions.

Because Richard was a screamer, the only people truly capable of withstanding the hurricane of his personality were extremely confident gay men. Richard's main assistant, personal assistant, and travel assistant were all buff gay dudes in their early twenties. Whenever Richard really, really cornered them, they fat-shamed him, and then, like a dragon under the influence of magic, he would go shuffling away to commit one of his stress activities like calling his on-demand chiropractic service or binge eating straight out of the aluminum trays from our catered company lunch. It was deafeningly satisfying stuff to watch.

Personal Assistant was the one who spent the most time on the phone dealing with Richard's gross health issues (and believe me, twenty-somethings do not understand why anyone would ever need to have a foot doctor and will absolutely share every detail of said phone calls with the entire office—or anyone on earth who asks), so his job wasn't great. But it was Travel Assistant who most frequently had to deal with the complicated demands of Richard's wife, whom we called Princess Smoothenface. Princess Smoothenface was an heiress by birth and a stay-at-home complainer by trade. Princess Smoothen was tall and blonde and always perfectly dressed, but she was also what is known in the scientific community as a "butterface"—absolutely beautiful until you had to look her in the eye. We suspected that her facial condition was what made her so impossible and mean. Thankfully, Travel Assistant was by far the most lacerating of our group, so he was more than equipped to handle her. Travel Assistant spent the bulk of his days on the phone with the American Express Black Card Centurion Concierge Service, arranging the couple's frequent "business" vacations in partnership with Deborah, Richard's longtime personal Black Card representative. By 2005 Deborah had been working with Richard's assistants for two years, so she was better at our jobs than we were and as fully in on the joke as a woman whose calls were "monitored for quality assurance" could possibly be. We loved Deborah. The guys constantly messed with her, hoping to someday crack her professional veneer.

"Yo, D-man, how's it hanging? Richard needs two first-class tickets to Florida this weekend."

"Where exactly would Richard and Amanda like to go in Florida?" Deborah asked cheerfully. The woman should have worked at the White House, she was so tactful.

"I don't know *where* in Florida, Deborah: Borelando? Tampon? I've only worked here three months, so you tell me where their horrible condo is. Excuse me, their *winter home*, as Richard just called it today, like it's a goddamned ice castle or something. Also, he wants to pay for the tickets in points, not money, because he's a dick."

Not paying for things was a standard request for the folks at the American Express Centurion Concierge. Working for Richard, I learned that the only thing rich people hate more than not having the best of everything is having to pay for anything. Nonetheless, Deborah did not hesitate. "We can definitely pay with points, but Mr. Jarron will have to incur a $30 processing fee."

"Wonderful! I hope he has a heart attack and dies in his office when he hears that news, Deborah. I hope you and I end up using these tickets ourselves to fly down to Florida while we skip his funeral."

"I'll have those ticket confirmations emailed to you within the next five minutes."

My conversations were with the assistants of our advertisers, so they were significantly less fun to listen in on than Travel Assistant's calls. At first the guys would make gagging sounds in the background, but after enough weeks of witnessing my anxiety, they simply took pity on me, pretending that I did not exist. Believe me, it was by far the more humane thing to do.

"This is Jess from Richard Jarron's office," I'd start, already stuttering and sweating in between my fingers. "I'm just calling to reconfirm our meeting tomorrow at 9:30 a.m. with Mr. Gottlieb, Mr. Marshall, and Mr. Johnson at the Regency Hotel?"

"Uh, yeah." Paper would shuffle and people would happily talk about other things in the background. "I think that one's still on."

"Well, can you tell me for sure?" I'd ask, smiling through my stress. "Because I need to know right now."

"Well, I don't know yet! I mean, God, hopefully something better comes up than breakfast with Richard Jarron, but if we have to take it, we'll take it. He's paying, right?"

"I can call back later if you need more time—how's thirty minutes from now? Will you know about breakfast by then? And also I don't have an email for Mr. Johnson in my system, and I was wondering if—"

"I've gotta go."

"Okay, but I still don't—"

"Bye!"

That would be Richard's cue to call up and ask whether I'd confirmed his meetings, most importantly his 9:30 a.m. with Mr. Gottlieb, Mr. Marshall, and Mr. Johnson at the Regency Hotel. The man had a gift for guessing exactly who had not yet confirmed. Any confusion about Richard's schedule would result in spitty, invective-laden screaming sessions delivered from the back of his custom-designed leather-interior van, a dimly lit business cave in which he frequently chain smoked Camel cigarettes. It was hard to know who was really failing at that point: him, the world, or me. By month two it had ceased to matter.

A couple of months after I'd started managing Richard's schedule I botched a meeting with his childhood hero, Phil Collins. It wasn't my fault: Mr. Collins, in the fine tradition of all rock stars, had dialed the entirely wrong magazine publication and told *their* publisher and CEO that he would be stopping by *their* office to say hello. One of the assistants sitting in that publisher's office had somehow been able to put two and two together and figure out what was going on—that Mr. Collins had intended to visit our shitty magazine, not theirs—and even emailed me at the last minute in hopes of remedying the situation. Unfortunately, I didn't see the email because I was busy organizing Richard's computer chargers. They were tangled and making him very unhappy, so to quell the tantrum I was sitting on the floor of his office, straightening and labeling each one. When we both figured out what I'd missed, Richard readied himself to shout, but Travel Assistant stood up and physically got in his way. "You have ZERO reasons to yell at her right now," Travel Assistant barked. "It is 1,000 percent not her fault your 1980s rock star friends don't understand concepts like calendars and telephones. What was she supposed to do—organize your computer cords with her own computer tied to her face? You didn't even buy her a laptop! Is she supposed check your phone and email every single second of every single day????!!! Because let me tell you, that is IMPOSSIBLE."

The next day an IT guy strolled over to my cubicle area as I was pushing my 2 p.m. salad into my face and enjoying my ritual afternoon check-in with the world's most important website, Gawker.com. The IT guy said that Richard had issued me my very own BlackBerry. My mood lifted. A BlackBerry! Only the most important executives in the company had BlackBerrys—only the most important executives in the *world* had BlackBerrys—and now I would be joining their ranks, at only twenty-four years old! In 2005 New York City there was a deep awareness of whose emails were stamped with the then-rare "Sent from my BlackBerry" footer and those whose emails were not. Sometimes people even faked it, typing in a "Sent from my BlackBerry" end to their note in hopes of impressing others, or to excuse grammatical mistakes, or simply to justify their own existence to themselves. But there was a discernible authenticity to the real thing—the specific way the BlackBerry font transmuted on different devices, the spacing of the stamp, the structuring of the email chain—a dozen tiny cues that email readers learned to pick up on as the technology proliferated. New Yorkers could detect the presence of a true BlackBerry with an animal sixth sense.

The IT guy had already set up my BlackBerry—the very latest and best model available. It was already vibrating and pinging away when he came over to my desk, happily doing its work with its Pavlovian little red light blinking. I was instantly enamored with the thing. Perhaps Richard was finally recognizing my talents and there would be more in store for me professionally than confirming his weekend brunch meeting with the head salesman of the Loro Piana in Manhasset. "We need a young mind like yours overseeing the magazine," I imagined Richard saying to me as he smoked his Camels, both of us grasping our respective BlackBerrys and hopping out of his business van. Walking alongside me down Avenue of the Americas in an Aaron Sorkin–like exchange about the future of the company, Richard would confess to me that he thought I "could really change things up." Quickly promoted, I would no longer need to check his calendar. In fact I would *be* one of his meetings.

"You need to keep your BlackBerry charged all weekend," the IT guy warned me, exempting himself from future claims of mismanagement and handing me multiple spare chargers. "Carrying a company BlackBerry means you're always reachable. There are no excuses for not returning Richard's emails. I just want to make sure I make you fully aware of that fact." I absentmindedly nodded as I pressed the device's tiny keys, watching as the software corrected my spelling mistakes and added periods to the ends of my sentences if I hit the space button two times.

"You've got to keep that with you at all times, even on the weekends," IT guy went on as he walked away. "In fact, because you work directly for Richard, *especially* the weekends."

Because I harbored none of the IT guy's scars or fears, my first weekend with my new BlackBerry was delightful. I did take the phone with me everywhere, using it on three separate occasions to call my parents from the street and feeling like I was gaming the system of life. I played Brick Breaker on the toilet like a real executive. I texted colleagues, yes, but also friends. I confirmed plans via email, a novelty back in the day. I added my more important friends from the magazine to my BlackBerry Messenger—when someone finally told me what BlackBerry Messenger was. Sure enough, on Saturday night my device pinged, bearing just one Richard email. For a second I panicked, but then I answered his question easily and went on about my evening. That night I went drinking at Down the Hatch with my Tisch graduate friends and listened to them complain about their latest auditions and the people they claimed were their agents. All the while my BlackBerry was quiet, but I was too new a user to be suspicious of its behavior. I assumed it was just another Sunday in Richard's life, that he was busy doing other things besides thinking about his stupid business.

On Monday morning I showed up extra early for work because I was a twenty-four-year-old with a jump on life. Over the weekend, after talking to my parents, I had resolved to take the job more seriously and continue my ascension by arriving first in the office. At 7:05 a.m. I started the office coffee machine, and by 7:08 a.m. I was seated in my cubicle. That's when things took a significant turn for the worse.

As I booted up my computer, my heart started to race upon my realization that Microsoft Outlook was thoroughly, nervously pooping its pants, overwhelmed by the task of loading the 346 brand-new unread emails in my queue. I'd had no idea, but my Blackberry had gone silent because the motherfucker had quietly broken down at some point on Saturday night. It turned out that the BlackBerrys of the day frequently liked to break down during the night and stop receiving emails altogether. They strongly preferred malfunctioning on warm weekends, and as a best practice they always issued exactly zero warnings of malfeasance to their owners when things really went septic. That was why I had missed so many emails. *It's not my fault*, I tried to justify to myself in my head. The mistake was BlackBerry's fault. Or maybe the IT guy's fault. Maybe he'd set it up wrong from the start, the idiot. But it didn't matter, because Richard would never, ever believe any of these excuses. Not in a thousand years.

Richard's emails were intimidating enough when responded to in real time, but they became completely unnavigable in large, unanswered groups. He always wrote emails in what Travel Assistant called "Privileged-Since-Birth White-Male Executive English." Spelling errors abounded, yes, but it was more than that. There were also cognitive-level errors bubbling underneath the more basic technical mistakes. If he was emailing about Ralph Lauren, for instance, he might really be emailing about Norwegian Cruises. Making matters worse, Richard also only ever used subject lines to write emails to assistants.

Sunday

5:37 a.m.

From: Richard Jarron

Subj: U need t call seema about the view

5:37 a.m.

From: Richard Jarron

Subj: Need amap of basel w clien names

5:42 a.m.

From: Richard Jarron

Subj: Call Smla us points for flights

Those Sunday emails, relatively calm and unconcerned in the Richard Jarron world, were the 5 a.m.-ers. By seven that morning, when Richard's needs had gone ignored for hours, the tone turned malicious.

At that point he was just going to start pig-piling needs on top of other needs for the sheer purpose of revenge.

7:31 a.m.
  From: Richard Jarron
  Subj: Called
7:32 a.m.
  From: Richard Jarron
  Subj: Called
7:33 a.m.
  From: Richard Jarron
  Subj: Ur vocemail is ful
7:39 a.m.
  From: Richard Jarron
  Subj: Where ar youvvvvvvvvvvvvvvvvvvvv

Why the Vs, you ask? Well! That's probably the question mark key pressed down without also pressing the shift key. If you are born rich, you do not need to press shift.

7:42 a.m.
  From: Richard Jarron
  Subj: is my 9:30 NOW
7:43 a.m.
  From: Richard Jarron
  Subj: vvvvvvvvvvvvvvvvvv

By 8 a.m. on Monday Princess Smoothenface was awake and finishing up with her Bikram class, emailing me with her own variation of exactly the same questions her husband had already asked. It was as if they never talked. They had been married for eight months. It was already the kind of marriage that seemed like it wouldn't last, but actually *might* last an affair-ridden lifetime because of money, familial expectation, and spite. "Spend more than two hours in a row with them, and you'll never believe in love again," Travel Assistant had warned me when I started.

Monday
8:04 a.m.
  From: Amanda Jarron
  Subj: Call me right now. How many miles points do we have? Richard needs to know.
8:05 a.m.
  From: Amanda Jarron
  Subj: Your voicemailbox is full

The emails went on in a similar fashion. Paralyzed and unable to stratify one email as more important than the next, I responded to the onslaught by texting one of my sympathetic, fascinated friends from the beauty editorial department about the situation. "Princess Smoothenface is emailing me too," Elisa texted me back. "She likes to take our extras from the beauty closet before the holidays for regifting. Even though, like, she's nowhere near as poor as we are."

I had to get to work decoding the messages. Desperate, I called Alice, Richard's ex-assistant, and I persuaded her to help me through the emails by promising her drinks and dinner on me. I knew the offer was tempting because she had walked out and quit on Richard without another job lined up. Plus, I'd have so much good gossip to share. Begrudgingly Alice agreed to help out, audibly satisfied at the idea of still being needed, and I began to forward her Richard's mountain of correspondence. That was the worst, most sadistic thing about being an abused assistant for too long: if you got good at it, you got addicted to it.

"Selma at American Express Centurion Concierge," she wrote me back immediately, "See how one time he spelled it 'Seema' and one time he spelled it 'Smla?' If you put those letters together and rearrange them, you almost get Selma. Selma is the old Deborah. So he meant to say 'call Deborah,' because he wants a map of the Baselworld watch festival, complete with a list of exhibitors by location. And Deborah may actually have that information, because she's amazing. Watchmakers are big advertisers in the magazine, presumably because English is not a first language for many of the Swiss."

Thanks to Alice, I now had a chance of surviving the day without getting fired. I called American Express and cried in loud, administrative tears to Deborah, who promised to make me her first priority of the morning. After answering my many questions about Basel, she politely commanded me to go to the bathroom and clean up. She knew what this Black Card–carrying assistant deal was like. She couldn't say so, but she was on the phone with the likes of me all day. I complied with her order.

I made it to the bathroom to clean up my mascara-drenched face. On the toilet I angrily confirmed the day's first two meetings with my now-semiworking BlackBerry. I called Wears-Sunglasses-Inside and screamed that today was 100 percent not the day to confuse Second Avenue with Second Street, and if he made any mistakes I would make the mistake of forgetting to retrieve his petty cash from finance every single day of the month. It was me vs. the broken BlackBerry, and by 8:08 a.m. I was almost-sort-of winning.

By the time Richard walked in the door I had thorough, perfect answers to his first six questions. "And what about my airline miles?" he then asked me, probably wanting to trip me up as punishment for the entire day of the weekend that I'd taken off. "Didn't my wife email you that I was wondering about how many airlines miles we have? Did you get back to her on that?"

Goddamnit, I had not. I wondered why on earth I'd bothered to pursue a college education at all.

I now understand exactly what the IT guy was trying to convey to me as I dicked around on my BlackBerry and ignored him: the little device he'd just bestowed upon me immediately redefined the relationships that we assistants had with our bosses because it transformed our jobs from jobs into very specific twenty-four-hour lifestyles. In becoming completely reachable by a piece of plastic and a onetime $400 investment, we were instantly repositioned in the eyes of our employers; we were no longer people you had to face on Monday morning to ask favors of; we were things you could text whenever you wanted. The expectations that the privileged started heaping upon their newly omni-accessible personal assistants quickly multiplied. If one rich person saw another rich person forcing their assistant to do something ridiculous, then that ridiculous thing immediately became acceptable for everyone to do too. I met assistants who fielded work texts while they were in labor. I met assistants who, on weekends, took their boss's children to visit colleges, see their therapists, or buy their prom dresses. I met assistants who spent half their lives mired in the paperwork of adopting a baby for their boss, only to then be tasked with assembling an elite team of round-the-clock nannies to raise it. I met assistants who worked for this woman, whose expectations were best described in the personal ad she penned for herself:

Can you find the right Gifted and Talented school for four-year-old Chinese girl? Can you find a holistic vet, a Japanese Hot Spa in Tokyo, and deliver a cut-bamboo arrangement in

*New Jersey? And get a corner table in a hot restaurant? No
Problem? Can you do it all in the same twelve-hour day? In
search of a world-class, extremely intelligent Personal Assistant
who thrives on stress.*

Almost all of the affluent people I encountered who were most comfortable with steamrolling over another person's private life were people
who'd experienced a life path like Richard's: people who'd inherited vast
sums of money at birth and therefore had never known what it felt like
to wait a table, or pick up someone else's phone, or be yelled at for
something like a thunderstorm. Outside Manhattan, New York City is
billed as the place for people who want to make it anywhere, but more
realistically it's the place for people whose families already *have* made
it everywhere, often for decades—sometimes for centuries. Although
I've long heard stories of some rough-and-tumble iteration of New York
City that existed in the 1970s, by the time I arrived in the 2000s the
city was good and cleaned up and a rich kid's kingdom, replete with
Hilton sisters. By the year 2000 New York was a place where there
were more applicants for private schools than there were available slots,
more $100-a-meal diners than $100-a-meal reservations, more Hermes
shoppers than Hermes handbags. By the year 2000 New York's famed
Condé Nast publications didn't have to bother interviewing debt-ridden
nobody NYU graduates like my friends and me because their intern
pool was already plenty well stocked with the children of clients and
friends. Condé Nast left the likes of us trapped somewhere in the portals
of Monster.com—résumés unscanned, references unquestioned. We
were destined to kick our careers off in less auspicious places.

One evening Richard announced that we four assistants were invited
to the magazine's latest issue launch party. The editorial team and

even the sales team always went to these fetes, and technically we *were* invited as well, but instead we often ended up anchored to the office late into the night trying to get a jump on the logistics of the next day. But on this occasion there had been clients in the office wondering aloud why we looked so glum and were left out of the merriment, and Richard blusteringly insisted that we forgo our after-hours responsibilities and join in. Travel Assistant and I freshened up and set off in a taxicab funded by our expense account. We looked up at the skyscrapers freshly glistening with whatever variant of DDT kept the city's cockroaches and rats from breeding or, god forbid, cross-breeding in Manhattan's most affluent neighborhoods. Of all the assistants, I got along the very best with Travel Assistant. He and I had that young New Yorker mindset in common: the belief that we were owed our small ration of the city's riches and were going to maintain a steady bad attitude until we got it. We'd brought roadies for the party commute and were bellowing along with the radio. We were secretly really excited to be attending. *So this is what city life can be like*, I thought to myself, a plump, gay Carrie Bradshaw who'd just gotten her first $250-limit Capital One credit card in the mail. *So this is what it feels like to have arrived.*

The location was a trendy downtown Italian place that had one floor for the tourists and a separate, secret, upper floor for the locals. Richard may have used the phrase "riff raff" when explaining the restaurant's social system's bottom class to us. That's the thing about New York: there is often quite literally a better version of the city hovering above the heads of the normal people. Inside Travel Assistant and I quickly found our outfits to be woefully inadequate. We were the only people not dripping in status symbols. My stolen beauty-closet makeup suddenly didn't feel like enough protective armor. I ran to the bathroom and cried for my circumstances within the privacy of a freshly painted stall. *You have nothing, and some people have everything!!!* I thought to myself. The emotion felt incredibly original.

Outside the bathroom I found Travel Assistant, who was making some of the same observations and complaining about his lack of an inheritance. This was the sort of topic on which we were able to dwell for days. "It's not my parents I'm mad at," he decided. "It's my grandparents. Do you know how easy it was to start a railroad?"

Of course we both failed to recognize that the Internet *was* the railroad and that in the early 2000s anyone could start a website. That's basically what Richard had done, though he'd transformed the website into a print magazine so he could charge a whole lot more for ads. Five minutes later Travel Assistant got a text from Richard himself on his phone. It read, *Can Jesus help me.*

"Party's over," Travel Assistant declared. "Because by this message, I am pretty sure that Richard is referring to you."

I leaned over and looked at the typo. Uncharacteristically I didn't even want Travel Assistant to plot a sarcastic, eviscerating response. I felt like the genie from *Aladdin* when it was summoned to grant wishes when in the middle of taking a bath.

"Let's quit," Travel Assistant said, picking up on the heaviness of my mood.

"Okay," I said, not really knowing what I'd agreed to.

"Let's go to Aruba," Travel Assistant proposed. "I just booked a fare there for Princess Smoothenface's parents. It was a dollar. I can put us on the same fucking flight."

Thanks to some very insider-y negotiations, the next day Travel Assistant and I were boarding in first class and bound for Aruba. It was the kind of flight where many people bring a pet in a box and clap for takeoffs and landings. The sky was bright and cloudless though somehow still proved to be an unmanageable roller coaster for Captain Tiffany, who hadn't made a single announcement after telling us about the *Star Wars* club she belonged to, the One World Alliance. Travel Assistant thumbed idly through a business magazine; I polished off my fourth mimosa. First class suited me, I felt. I'd grown up with parents who were deeply disdainful of people who paid to sit in first class. *Yeah, but what a way to go,* my nine-year-old self had thought back, envious. I would undoubtedly be wearing all my snap bracelets in order to seem more glamorous than the rest of my family. As I stared dramatically out at the clouds I wondered what those first-class passengers of yesteryear would think of me now, finally one of their own.

I woke up as the plane was descending. The flight touched down to a storm of applause. From somewhere in the back of the cabin a chicken clucked in excitement. It was around this time that we discovered that our BlackBerrys had not yet been shut off by the company—in fact, we had multiple messages awaiting us. Nice messages. Then the calls started coming from the magazine's human resources department. Could we please call back as soon as it was possible? Everyone was very concerned. We couldn't just quit! More like holy shit, we couldn't *both* just quit! Richard "needed" us! We saw how this sudden unexpected onset of concern and respect was working. With us in Aruba, who would confirm the week's meetings? Who could finalize the week's travel? Who would talk to Princess Smoothenface when she called with her own list of needless, distracting demands? Thinking on our feet, we negotiated a deal where we'd be paid for two more weeks, but only work for one. This, we believed, was the most advantageous professional parting arrangement that anyone had negotiated in the history of commerce. In exchange for the money, the company would avoid Richard missing a meeting with any of the most important heavy hitters in his Rolodex, such as Phil Collins, while the remaining members of the HR staff scrambled to locate and train some more assistants.

Travel Assistant and I spent three days on the beach ordering a constant flow of deep-fried appetizers. We talked of our futures for hours from the warmth of two feet of Caribbean water. We met a lizard. We got sun poisoning. We drank so much we thought we saw Puff Daddy. And yes, thanks to the life-changing wonders of technology, in between the highlights of our vacation we dealt with Richard's flights, emails, and calendar confirmations with exactly as much concern as was warranted from someone making far less than $10 an hour: practically none.

The Gawker
New Yorker

What is the point [of Gawker]?
To say what other people can't. That's the fun of it.
—Elizabeth Spiers, 2003

W HEN I first moved to New York and began reading the city's celebrated, eponymous periodicals, I started wondering whether my colorful new home might actually be insane.

It was the *New York Times'* Style section that most reliably reinforced this hypothesis. Each week, regardless of war or famine, the Style section prattled on about topics like retailers trained to remove the troublesome scuff marks from valuable vintage sneakers. *Do New Yorkers realize that other people think it's totally ridiculous to pay $500 for a pair of plastic shoes, and then pay more money for someone to clean them?* I wondered to myself, totally baffled. I had understood the *New York Times* to be the paper of the people, and I had figured that if said paper was going to recommend some footwear, it might cost $20. Clearly the New York papers' austere advertisements, stupefying real estate listings, and overall calm regarding the stasis of inequality in our city didn't affect other New Yorkers as it did me; clearly they weren't as upset as I was. Sure enough, by year two I stopped remembering to be surprised—in fact,

I even began looking forward to my weekly trip to the Style section. It felt sort of like visiting my friend's rich aunt in New Orleans who'd once explained to us, in all her pancaked-makeup glory, that the best way to lose weight was "One Lean Cuisine a day and vodka."

During my midtwenties I was working my terrible job at my terrible, advertorial-laden magazine and always secretly attempting to get another job, any job, at another magazine, preferably one where I could write about computers. My most promising run was a string of interviews over at *Fast Company*, where I was hoping to become the assistant to one of the editors. By the third interview I was getting extremely pumped up about my prospects, walking to their office and blasting my iPod like Dwight Schrute before a sales call. To prove the depth of my interest in the position, I even showed up to that final interview with a manila folder full of my own article ideas on various technology companies. I did not get the job. "It seems like you'd be unhappy as an assistant, and you'd want to do more than answer phones and get coffee," the editor told me when I didn't ask for an explanation as to why I didn't get the position. *Yeah, well, screw you, alacrity*, I thought to myself.

Chained to *Craptacular* magazine for the foreseeable future, one morning I opened a 5 a.m. email from our publisher, Richard, that would eventually help me figure out how New York City life really worked. "WE HAVE A NEW INTERN STARTING THS WEEK, SO NBODY MAKE HIM DO ANYTHING HE DOESN'T WANT DO!!!! HE IS THE SON OF A VRY IMPAOTNT CLIENT!!!!" Richard bleated, with all his usual sent-from-the-treadmill BlackBerrying vim. "How dumb is that last email from Richard," I texted a couple of my friends at the magazine immediately after the intern decree went out. "What kind of idiot takes an unpaid job at our stupid magazine?" I felt hugely superior to this misguided fellow twenty-something. Until I heard back from my friends, that is.

"Uh, Jess, these interns are usually the kind of 'idiots' who don't ask for a salary because they don't need one. They just need to say they had an internship so their parents can get them a job at a better place," my friend in editorial quickly texted me back. "Kiss their asses," she continued. "You are sad to them."

Sure enough, Intern Kid arrived the next day with his wrist already Rolexed. He would never be screamed at by Richard, not even for the most egregious of violations on Richard's list of shout-worthy workplace crimes, such as not knowing who was calling on another employee's phone line. Instead, Intern Kid would accompany Richard to meetings with record label executives and Fortune 500 CMOs and be invited to attend lengthy client lunches at Cipriani or movie premieres at the Ziegfeld. After these experiences were ingested, Intern Kid's opinions about how to better the magazine would be listened to and some of his suggestions even heeded ("We should have a monthly column about where to buy the best snowboarding gear in Tahoe"). In a month or two each Intern Kid would head off to some other, better part of the publishing industry, résumé now clad with enough "real-world experience" that he or she was deemed fit for a low-level editorial job at a magazine that didn't outline its fact-checking process as "Call all the advertisers and make sure they're happy." It was the first time in my life I watched someone younger than me professionally outpace me. It would not be the last.

What is Gawker? Current obsessions include but are not limited to, Tina Brown, urban dating rituals, Condé Nastiness, movie grosses, Hamptons gauche, real estate porn, Harvey Weinstein, fantasy skyscrapers, downwardly mobile i-bankers, Eurotrash, extreme sport social climbing, pomp, circumstance, and other matters of weighty import.

—Gawker, 2002

Anger can be a very productive emotion. Gawker was a blog born from a small explosion of urban twenty-something agita. Like me, Gawker's writers were new New Yorkers who'd arrived in the city high on the self-esteem parenting of the 1990s only to shit their pants at the inequality abounding. *But you don't understand*, we thought to ourselves, *back home people thought I was talented.* And also like me, Gawker writers were people ignored by important portions of the city's publishing world in part because their families lacked publishing world connections.

Locked out of the traditional media universe, Gawker created the untraditional media universe in which it would reside for the next ten-plus years. Here we had a microcosm of the real two-party political system of Manhattan: the haves and the have-nots, the insiders and the outsiders. Gawker, run by a motley assortment of defiant have-nots, was very invested in using a different kind of scorecard to chart the things that happened in our city. Soon they were loudly chastising the choices of institutions such as the Style section, like a teenager criticizing their mother's nightly glass or three of wine. As part of such investigatory efforts, Gawker used to pick apart the wedding announcements in the *Times* in a regular feature they'd entitled "Scoring Sunday's Nuptials." Each week Gawker writers plainly added points for all the life advantages that had been handed down to the bride and groom by circumstance prior to receiving their press coverage in the *Times*, for example:

Katherine graduated from Princeton: +3
Katherine graduated cum laude: +1
Katherine's grandmother, Countess Alicia Spaulding Paolozzi,
helped Gian Carlo Menotti found the Spoleto Festival U.S.A.
in Charleston, S.C., in 1977. She also drove for the winning
women's team and is the accessories editor at Harper's Bazaar
(media job): +1

Andy was an editor at the Paris Review, now an author
(media job): +1
Andy's parents are from New York: +1
Katherine's mother is a trustee of the Contemporary Art
Museum in Houston: +1
Andy's mother was a model, and there's a nude picture of her
by Richard Avedon at the Met: +3
Andy's maternal grandfather was in the 1958 automotive Tour
De France: +3
Andy is thirty-eight:-1

A Gawker writer who referred to herself to as "Intern Alexis" had constructed a "complex algorithm of patrician lineages" that helped her and other staffers score these power-nuptial announcements Sunday after Sunday. Consistency was key, as every week there was always a loser and a winner in the wedding section analysis, but as is true in New York, one plus one did not always equal two. A sampling of the Intern Alexis algorithm:

Investment banker: 2
Both Investment bankers: 5
Management consultant: 1
Both management consultants: 3
If bride or groom attended or taught at any school with
"Country Day" in the name: +2
Mother a kindergarten teacher or reading specialist/father a
wealthy industrialist: +3

It was my magazine job and my Gawker habit that first illuminated to me how the system of New York worked. *What is this island?* I'd wondered to myself. I now understood that there were systems that kept the powerful in power here, and although there was lot more to it than Intern Alexis would ever be able to calculate in listicle form, she wasn't all that far off. This is when I began to see the true purpose of Gawker's writers when they were working at their full potential.

Adding to the general digital muckraking merriment, Gawker not only wrote for us nobodies but reported through us as well. Gawker used the immediacy of the web to build an underground army of interns, assistants, copy editors, waiters, and other hooligan informants: young people placed all around the city who were always willing to blab in real time. Was there a well-known media boss at Michael's restaurant, screaming at their employees for ordering smelly, spicy foods in front of a client? Gawker wanted in on that information, and they had a well-engineered system to retrieve and publish it within the day.

Fashion's Best and Worst Bosses:
Beneath Donatella's tanned cowhide exterior lies a gooey
Mamma Mia center.
But watch how you make the coffee for Calvin—match it to
Pantone chip 499, please, or start brewing another pot.

—Gawker, 2003

Gawker quickly became the personal assistant's daily paper of record, the new brand of online text opiate for the urban, under-careered, twenty-something masses. Gawker's first editor, Elizabeth Spiers, was a funnier, craftier version of all of us: she was a kid with good grades who wanted a seat at a prestigious magazine, but she hadn't quite gotten the gig. Unlike the rest of us, Elizabeth set out to make her own party, laughing at big media instead of complaining about it all the time. That was the example that Elizabeth set as she sat down every day to write for Gawker: she made it cool to be uncool. We anointed Elizabeth our leader because she'd reacted to the city's disinterested bitch-slap with verve.

The thing we young readers loved most about Gawker and Elizabeth was that Elizabeth coveted the exact same "media elite" jobs that she routinely mocked. Elizabeth and crew stalked the comings and goings of people like Anna Wintour because they were as jealous as they were admiring. Many traditional media folk tried to wash their hands of the Gawker narrative by calling the notion of both teasing and idolizing the same thing a disingenuous tale. To us Gawker readers, however, it was the only truth we'd known. We were both envious of and angry at our peers over at the better publications and occupying the better media jobs. We liked making fun of the goings-on at Condé Nast as much as we would have liked to work there. It simply was.

There were many assistants at the magazine I worked at—more than a dozen lunch-fetching grunts and phone grunts and Excel grunts and executive-specific grunts—and we all checked Gawker several times a day. We worked for a well-known and terrible man who'd married exactly the woman he deserved, and Gawker enthusiastically crucified both of them on a number of occasions. Gawker writers fluently spoke the language of Personal Assistant—they knew what it was to be the one who scraped someone else's guacamole throw up off the company toilet because "clients were coming in"—and on the days when we opened up Gawker to see our own horrible boss and his horrible wife oh-so-accurately described and destroyed, we laughed from a place in our souls that I don't think I even have access to anymore, now that I'm older, now that I'm more comfortable with myself, now that nobody yells at me when there's traffic. We felt less insignificant. What would have been a story we'd told a couple of disbelieving friends at a bar had now become permanent, searchable testimony.

As the 2000s continued and the whole "Internet trend" failed to fade into the background as so many a media executive had hoped, magazines were forced to contend with a shrinking and shape-shifting business model. That was when Gawker really kicked into gear. The blog was growing up. Many of the major publications that Gawker spent its days criticizing had now taken the time to profile the blog, its writers, and their purpose. The big media world was suddenly less intimidating to Internet writers. And hell, maybe the Gawker writers' and Condé writers' fortunes would someday reverse and the Internet would be the source of riches. It was a theory worthy of at least a pause and a shrug.

More confident in its right to have a seat at the media table, Gawker became a spritely frenemy to the powerful, strategically picking its fights. Gawker began drawing attention to the Elagabalus-style personalities atop the mastheads of several major publications. The writers had a special love of mocking Graydon Carter at *Vanity Fair*, as *Vanity Fair* was a magazine unafraid of ridiculous spectacle, even as that Rome of traditional print media was partially burning. Regarding the rumor that Mr. Carter had banned the use of the word "chuckled" inside his magazine's pages, Gawker wrote,

I use the word "chuckled" because existing assets report [to me] that "chuckled" has been banned from the Vanity Fair vocabulary repertoire. Curiously, the following words have not been banned, as of the April issue: "luxuriant," "Liza Minelli," "exquisite," "nibbling." Astoundingly, the following sentence has not been banned: "A few years ago, I sailed into the harbor at Cap d'Antibes aboard a friend's boat."

—Elizabeth Spiers

One of *New York Magazine*'s annual traditions has long been the very casually bourgeois "100 Reasons to Love New York" listicle that its writers pen around every Christmas. In 2005 that list included such sincere reasons to love the city as "Because You Can Get a Nice Plate of American Hackleback Sturgeon Roe at 3 a.m." and "Because the Fountain at the Brooklyn Museum Makes Music." Gawker, smelling formulaic rich people tiredness, saw an opportunity and compiled its own list of reasons to love New York. The Gawker list included such bon mots as "Because we can get drugs delivered to our doors," "Because you can take a dump at the Apple Store," and "Because we get to push tourists if we're late for work."

By 2007 Gawker had long loved any screamingly tone-deaf managerial decision made by the magazine industry or any general underestimation of the Internet as force for change. They would soon be handed the Holy Grail. This story would be one of the first multiyear stories in which the Gawker perspective would not only be unique and valuable but also ultimately right. Condé Nast had announced that it would be investing a sizable amount of capital into a new magazine called *Portfolio*.

Portfolio was a bet-the-house attempt at saving Condé Nast from the repercussions of the Internet world by doubling down on the print world that had come before it. With $120 million-plus in launch budget, the project was a classic "too big to fail" business strategy. What was this new magazine's purpose? *Portfolio* would pair the approachable glossiness of a woman's fashion magazine with the important big-thinking-ness of a man's business magazine. As reported by *New York Magazine*, "The premiere issue promised ad buyers an average reader who's a wealthy, 42-year-old male, and claimed there's a 30 percent chance he's a "C-suite" executive (i.e., CFO, CEO, or COO)."

Millions of readers, millions of them male CEOs. These assumptions were so mathematically delusional that they deserved to have a bunch of twenty-somethings in sweatpants mock them. Gawker pluck-ily, publicly placed its bet, stating that Condé Nast and *Portfolio* were going to fail. Big time.

You can guess the end of this story: even with more than a $100 million to spend, there isn't really a way to reinvent the magazine. You can't make the articles dance. People can't hop inside the pages and travel through time. It's just a magazine, and just like this book, one of its primary jobs is to look nice on the top of toilets and effectively collect condensation. After Gawker had made its prediction of doom, eventually the gossip started rolling on in: *Portfolio* was in chaos. The publication had apparently spent its inaugural summer doling out free lunches and accomplishing little else. Feigning cool, Gawker released word of these rumblings to its readers and immediately rechristened the publication *Fort Polio* magazine. It was now open season.

Fort Polio's fall was the Internet's win. A little blog that ran like a rusty old Ford Taurus with its bumpers duck taped together was now chronicling the takedown of a bunch of adults in a glamorous high-rise tower eating a continual free lunch buffet as they purported to be changing the magazine industry forever. This was a bigger, better use of the patented Gawker anger that had fueled its earliest chronicling of the elite.

Over the past fifteen years there have been many enjoyable moments in the big media–little media dialogue. As I've gotten older I've real-ized that it's not always one group versus the other, that the *Times*, the *New Yorker*, and our other papers of the city are, to varying extents, in on their own joke when it comes to some of the rich-people cultural commentary they provide. One of Gawker's own, Max Read, wrote of this phenomenon:

See, the secret of the Style section is that it's intended for two
audiences. The first audience is its "official," explicit audience:
people who see nothing problematic with being told by The
New York Times what's cool, and think of the Style section is a
straight-ahead, un-ironic record of hip trends and cool people.

"But there is the second audience," Mr. Read continues,
finally explaining why my subconscious felt so irked for so
many years. "A secret audience. . . . This audience reads the
Style section, week after week, and thinks "what the fuck is
wrong with rich people?" This audience regards the Style
section as a collection of dispatches from a different universe;
a universe where some of the most horrible and insufferable
people on the planet are treated as visionaries and geniuses."

The Times is aware of this.

Are the rich people featured in the *Times'* Style section something
to admire? Mostly no. But they are something that exists, and their
existence is absolutely news that's fit to print. Their power takes a long
while to understand, as it cannot be covered in a 750-word piece—or
even two years' worth of 750-word pieces. Understanding their power
is in the details. It's hard to report neutrally on logic that is Boolean:
not all people who buy $500 shoes have spent their lives systematically
oppressing other people, but many oppressors do buy $500 shoes. This
is why the *Times* covers the city's many important rich people as carefully and calmly as it can. It's not as simple as good vs. evil.

Gawker and its copycat spawn have never been a perfect medium of
dissent. After Hurricane Katrina, in what was arguably the most brilliant simultaneous use of live television and the then-nascent YouTube
as mediums, a calm and heartbroken Kanye West took to the national
stage and famously declared that "George Bush doesn't care about

black people." In totality Mr. West's comment would later be proven so percipient that George W. Bush would admit that it represented the low point of his presidency.

Gawker, of all outlets, should have been celebrating Kanye West's candor, his refusal to stick to the approved mass media message when human lives were at stake. Had Gawker spotted the truth in those words, perhaps someone could have won a Pulitzer. Instead, the blog loudly missed the point, describing Mr. West's rant as "fantastically batshit." Rather than researching the photographs that had partially inspired Mr. West to comment on how the press at the time[1] was portraying African Americans, Gawker preemptively dismissed Mr. West's allusions. In a moment of inexcusable ignorance, acting exactly like the glib, bratty teenager texting at the funeral that Gawker's enemies loved to accuse it of being, Gawker prattled on, calling Kanye West "blessedly inarticulate." Comparatively, the *Times* seemed to more calmly see the point: "Gulf Coast Isn't the Only Thing Left in Tatters; Bush's Status with Blacks Takes Hit," they wrote in covering Kanye West's words. Sometimes avoiding extremes can help a group of reporters to start to see the truth.

There are, however, important instances in which I believe that Gawker was clearly in the right and the traditional media world was clearly in the wrong. One of them involves the Walton family, the living heirs to the Sam Walton of Walmart fortune. Hoarders of money, dodgers of taxes, employers of tacky florescent doom—to me the Waltons are the kind of people who render praise impossible. That's why I was thrilled

---

1. Notably the AP wrote, "A young man walks through chest-deep flood water after looting a grocery store" in reference to a photograph of a black man, while the AFP wrote of a photograph featuring a white man and white woman in a near-identical pose: "Two residents wade through chest-deep water after finding bread and soda from a local grocery store."

one day to find that one of Gawker's all-time best writers, Hamilton Nolan, had an interest that was very worthy of the blog's by-then considerable cultural cache: Alice Walton had announced that her great act of charity to the world would be to open an art museum. This charitable museum would be funded with the money she'd inherited by birth, saved by not paying taxes, and stolen by offering slave wages to her father's company's employees. Appropriately incensed, Mr. Nolan wrote,

> *Four members of the Walton family, heirs to Sam Walton's Wal-Mart fortune, are collectively worth more than $100 billion— more wealth than the entire bottom 40% of Americans. . . .*
>
> *[The Waltons] have decided to hoard as much of their fortunes as possible. They have decided to use each and every tax loophole possible in order to keep their money in their own family, and not to allow the public to claim a single dollar more in taxes than they absolutely have to. . . .*
>
> *So the next time you hear about how fabulous the Walton family's opulent new art museum is, remember that all of that charity is part and parcel of a structure designed expressly to hoard billions of dollars within this one single family, and to avoid paying the normal tax rates. The Walton family's very existence is an insult to the American dream.[2]*

Now let's turn our attention to the *New York Times*. What did they have to say about the Alice Walton and her art museum?

> *"For years I've been thinking about what we could do as a*

2. Hamilton Nolan, "The Waltons Are the Greediest Family in the World," *Gawker,* September 12, 2013, http://gawker.com/ the-waltons-are-the-greediest-family-in-the-world-1300311273.

family that could really make a difference in this part of the world," said Ms. Walton, who is 61. "I thought this is something we desperately need, and what a difference it would have made were it here when I was growing up." . . . A celebrated horse-woman who seems more comfortable in sneakers than stilettos, she is the youngest of four children; while her eldest brother, S. Robson Walton, is chairman of Wal-Mart.[3]

If that is a joke—an instance of the *Times* using subtle literary tools to surreptitiously mock the rich—then no one is laughing. The *Times* has covered or blurbed Ms. Walton's Crystal Bridges Art Museum in a positive light no fewer than nine times—so many times that it makes the museum's existence feel normal. What we are witnessing in the instance of the Walton family isn't some editor's shrug of their shoulders, some acquiescence to the existence of things like $500 shoes. What we are witnessing here is our nation's paper of record allowing a despot to hide in plain sight.

Unlike the *New York Times*, Gawker never had anything nice to say about the Waltons; Gawker relentlessly held them to the fire. Gawker never praised a robber baron's fucktastical art museum because it's the robber baron part that's important, not the art. The question Gawker seemed to have for the *Times* was so simple that it almost felt unanswerable: Why write something nice about a Walton? Who benefits? Certainly not you, *Times*: remember the math lesson of *Portfolio* magazine, that there is no huge reader base of billionaires. It's best to write for the rest of us.

Last year a dangerous blowhard's lawsuit dismantled Gawker, and the Internet mourned the loss of one of its most original gathering

3. Carol Vogel, *"A Billionaire's Eye for Art Is to Shape Her Singular Museum,"* New York Times, June 16, 2011, www.nytimes.com/2011/06/17/arts/design/alice-walton-on-her-crystal-bridges-museum-of-american-art.html.

100

places. The young people's New York City decoder ring was shuttered, though its record reminds-with all its shaming of the Waltons, tailing of the magazine editors with the $100 hundred expense accounts, and accounting of the world's worst bosses. Where do the city's twenty-something residents go to complain now? Fucking Snapchat? That thing is *idiocracy* brought to life. The only pursuit less intellectual than staring at your own face is staring at someone else staring at their own face. Please.

I don't know if I would have made it through my New York twenties without Gawker. It was the little guy's voice, now sadly missed in a city where numerous annual "Fashion Weeks" create reams of news and Waltons are allowed to roam free. It was something important.

2005-2010 TECHNOLOGY, REMEMBERED

Myspace is now the place where you go to listen to new music from your favorite band, the local one trying to cultivate a following on Myspace without involving The Man or any of his global conglomerate recording companies. Thanks to Myspace, we all sort of start to subconsciously realize that sometimes The Man actually is helpful because he makes sure things are good before us general public have to subject our ears to them. He makes sure the work of Myspace musicians such as Tila Tequila stay relatively contained. Perhaps The Man is doing God's work.

Twitter emerges. Sharing your thoughts, original and unoriginal, becomes the new medium of communication for the Internet. "Getting coffee." "@The Robinsons." "YOLO: 'carpe diem' for stupid people."

Comedians, seeing how bad regular Americans are at one-liners, appear and take the whole medium over. A couple of people get book deals.

Three months in, all the other comedians become desperate. Where are their book deals?! Twitter is a place for having the best jokes! Everyone must have to the best joke first! "First" means milliseconds after whatever the event being joked about happened. Twitter reveals its true potential to become a smiling, waving, self-promoting, screaming digital hellscape. Desperation abounds, and everyone feels they deserve and need to have thousands of followers to survive.

While the Great Comedians War of the Twitter 2000s–2010s rages, brands also join the party. Assuming the casual language of personhood, they hugely embarrass us all with their attempts at fitting in . . .

Why I Hate Twitter, The Publicly Traded Media Company

I F you are young, live in New York, and have more than seven followers on Twitter, congratulations: you are a social media expert. You can get a job pretending to be a prominent business person or even a publicly traded company on Twitter.

Since approximately day ten of its life, Twitter's bread and butter has been selling its magical, invisible Twitter "influence" to the Fortune 500 companies of the world. Technology journalists helped create this monster when they wrote story after story in the mid-2000s about how great Twitter was because now people could "interact" with all their "favorite companies." "Twitter is a democracy!" everyone shouted. And it turns out, they were sort of right. Twitter is some mutant, heinous form of a digital democracy: it is a publishing tool that gives a voice to the truthful and the lying, the paid and the unpaid, the human and the robot, the grandma, and the global corporation all at once. Twitter is the democracy truly designed for the Citizens United world. Twitter is the Internet's original "Let's all hang out and talk to each other" social experiment, now being sponsored by Wendy's.

I was in my midtwenties when Twitter came into its power. As a computer-loving, air-breathing young person, I possessed all the qualifications necessary to work at a social media advertising agency and help corporate America's companies "get online" and "become relevant" to our nation's digitally empowered young people. In return for my services, I would be paid almost fifty thousand dollars a year. For at least a day after I signed the paperwork for this job I was living in geek heaven. I would get paid a ton of money to spend quality time with my computer! I liked social media more or less the same as anyone my age. This job would be great! Oh, how wrong I was.

Social media is a great business for adults who happen to be emotional children. When you go through the doors of the average social media agency, it looks as if you are walking into the daycare center at the Crayola factory's postapocalyptic explosion. Saturated, exuberant colors bellow "You are having a meeting here! YOU! With really CREATIVE PEOPLE!" and in the waiting area things that are usually made out of furniture are now made out of foam. Many well-compensated employees do not wear suits to work: they wear imported jeans that are too big for them, falling off their aged butts. In social media agencies there are walls made entirely out of white board material where random people who make tweets for mayonnaise can just potentially "be creative" at a moment's notice. Say you're a middle-aged creative director exiting the bathroom and haphazardly cleaning your hands with a worm of Purell. Midsanitization process you think up the perfect tweet for a new Folger's coffee, one targeting the confident, health-minded moms who value time with their families and make up your core demographic. Well, now, thanks to the walls that are made of whiteboard material, you can just walk over to any surface and sketch out your idea right there! This sort of spontaneous gesture never happens.

One of the most successful senior-level executives I first met within the first week at my social media job was a man who worked entirely at the speed of acronyms. "The SVP of MGM is in LA?" is something I sincerely heard him inquire on a conference call. This gentleman wore exuberant, paisley-patterned shirts and took sloppy notes on mathematical-looking pads of graph paper during meetings. One day he explained his sartorial choices to me by clarifying that crafting truly excellent social media was "both an art and a science." Hence, the flowers and the graph paper. He had two summer houses.

Back in the days when traditional print and television advertising were the only games in town—back in the glory era of smoking camels, car insurance geckos, and frosted cereal tigers—all you had to do was give an animal a hobby and you could and make a mint in the business. Then Twitter came along, and suddenly some brands started using the shiny new mobile medium to "talk" with their consumers one-on-one, as if the brand was a person too. For some reason corporate America lost its ever-loving mind. All of a sudden it was expected that brands would *listen* and *talk* to the Average Joes and Janes who bought their products every single second of the day, not just shout at them before movies or in between TV shows. It was expected that brands would quickly respond to people's complaints, ideas, and misanthropy; their criticisms, suggestions, and their misinterpretations of basic information. (This was no small task: a beauty company executive once told me that every year their 1–800 number has to handle a certain volume of calls of people claiming their razor "doesn't work." Apparently the correct first response from the customer service rep is "Did you take the cap off?" It is hard to speak calmly with America.)

Due to the understandable overwhelm, social media instantly became a thing that big companies outsourced, a category that "experts" could sell their "expertise on" to the major brands of the world. Thanks to these people, the brand known as Kit-Kat has over eight hundred thousand "followers." At the time of print Kit-Kat's Twitter profile reads "KITKAT—the perfect treat with your tweet. #HaveABreak." As a former industry insider, I can assure you that at least one hundred people were involved with writing that sentence and hashtag.

This is their story.

There were two groups of people who worked at our social media agency: the older people, who lent their client-management skills and résumés full of advertising awards to the firm, and the younger people, who made the social media. The older folks were in the twilights of their careers, the flush advertising industry having provided a very good life for them and their families. Multiple people owned boats. As is typical in a workplace, the younger people resented the older people for their financial comfort and for doing what seemed like less work, and the older people resented the younger people for being obnoxious, high-maintenance nightmares. Each group had their points. Thankfully, the two factions really only spent time together if a client was in the room, which kept everyone on their best behavior.

Regardless of whose narrative was more true, the old people's or the young people's, the twenty-something staffers in our social media agency immediately bonded together like a tribe of Patti Hearsts. We gathered nightly in bars to recuperate from the stress of crafting the corporate tweet. We constantly spoke of quitting our jobs and moving upstate to, like, start a farm or sell blue jeans or literally anything else. We hated social media, but our jobs were what all Manhattan jobs are to all the city's twenty-somethings: a way to not go home and admit to everyone that you hadn't made it. We were learning the hard way that in New York success was often just a different way of failing.

Toward the tail end of my first month we learned that our agency was going to pitch an important new social media account. A major CEO apparently wanted his company to "get on Twitter," and we were one of the agencies he was considering for the job. To the modern, time-stretched, and digitally unaware executive, "getting on Twitter" was never as simple as merely typing in the URL into a computer and typing a fucking tweet. For the men of these corporations it meant a multi-million-dollar deal with a social media advertising agency. The company we were pitching was a global conglomerate with more tentacles than most *New York Times*–reading citizens could ever track, let alone understand. Through careful research we learned that our potential client was responsible for the manufacturing and marketing of a wild array of consumer and commercial products, including but not limited to frozen family-style dinners, affordable pet food, farming pesticides, imitation wood polymers, cigarettes, and medicine. A perfectly functioning circle of modern life.

Our boss brilliantly mitigated the awkwardness of a business meeting spent talking about "the always-on Millennial generation" and "social media" by providing a lot of top-quality snacks for guests to eat. She knew that the conversation was going to be grating, and it was best to keep everyone as thoroughly distracted from tracking the content as possible. Appropriately, there was a three-inch-thick office guidebook for the office assistants that explained how to prepare all the most important attention-destroying business appetizers, such as the Individual Cheese and Fruit Tray, the Dueling Bowls with Original and Peanut M&Ms, the Mid-Meeting Warm Salted Nuts Surprise, and the End-of-Meeting Bite-Sized French Dessert. Every single obstacle that could ever prevent a grown person wearing a $3,000 outfit from publicly stuffing their face had been thoroughly considered and eliminated. Why the individual cheese tray? Because adults don't like reaching over other adults for snacks. Why the addition of fruit? Adults all pretend to be healthy eaters in front of other adults, even though we are a nation that lives under the totalitarian regime of the heart attack. When it came to the beverage portion of the session, our boss preferred setting afternoon meeting times because she was a big believer in serving wine. She even had waiters on staff full time who automatically poured the wine for our guests, guessing their preference by way of an ironclad flowchart designed by none other than the boss herself: men liked red wine, and interesting women liked red wine, and uninteresting women liked white wine. This rule probably poured a less-than-5 percent rejection rate.

In retrospect I can say that had there been any lesser amount of drinks and snacks at this particular business meeting, the group would have succumbed to the numerous constraints of discussing something that most people at the table didn't like or understand—social media—and our agency never would have gotten any money.

As our prospective clients made their way into the conference room and the snacks were set out, we young people suddenly became nervous. It was weird to look around and see this emotion in each other—genuine concern—as our plan had always been to never care about anything work related. Suddenly we realized that we wanted these clients to like us. We wanted to seem cool to them and smart about computers. We didn't want some other kids at some other stupid social media agency to seem cooler than us, as by our calculations that would have been a double life insult. The elder executives were comfortably in their business element once the pitch started, mucking it up with the clients over kid-related summer camp bullshit and the transportation nuances of a summer's worth of vacation plans. We Patti Hearsts didn't have any vacation plans. With any luck we were still going to be here, in New York City, making social media. We sat and we waited. Our prospective clients, midwesterners in practical shoes who tolerated the chicanery of the social media industry with Pepto-Bismol grimaces, listened politely while we agency folk opened the meeting by introducing ourselves in a deeply over-rehearsed round-robin of statements: quick monologues that included our names, titles, and one pre-approved "interesting fact" about ourselves. The client team smiled blandly. They then introduced themselves with their titles only, forgoing the humiliation of interesting facts. Our boss pressed a button on the small remote she kept discreetly in her lap, and approximately one minute later a platter of warm chocolate chip cookies arrived. Cookies! Even business midwesterners experience joy at the sight and smell of cookies. The meeting was saved, at least for a moment. The next morning our boss would issue a company-wide email with the subject line "Best Practices," decreeing that no one at our company should ever kick off a client meeting with interesting facts. "Clients want to get to the point. They're not here for idle chit chat," she wrote. Of course, there was no mention of the fact that including interesting facts was the boss's idea, and confusingly, the clients *were* here for idle chit-chat, because that's what Twitter was.

We quickly moved into the presentation portion of our meeting, during which I was scripted to address the idea of us kids ghostwriting for this prominent CEO on Twitter. The topic of me tweeting for him felt about as comfortable as discussing sex.

"Let's talk about what kinds of things you'd like to say on Twitter," I posited faux-casually, opening the conversation exactly as it had been scripted and almost shitting my pants.

"Should I be on Twitter myself, or just the company? What should I say if I go on Twitter?" the CEO asked all of us young people, attempting to sound like Socrates but instead sounding more like someone who was pissed off that this whole stupid problem of online-transparency-mobile-first social-media bullshit wasn't already solved for him.

"It might be fun for you to go on Twitter. Do you have any hobbies?" I asked the CEO in return, skipping ahead to question six out of blinding nervousness.

"You think I have time for hobbies!" He laughed the way a monkey in a zoo laughs right before pelting its keeper with bananas. Here I was, a useless twenty-something who thought business people became wildly successful because they devoted a lot of time to their hobbies. Just because President Obama liked to dribble a basketball around didn't mean that *this* man could run a pet food, people food, environmental toxin, and medicine-making company all while simultaneously cultivating some suite of hobbies.

Thankfully a different young person began speaking—a kid named Winston who had been in a new business pitch before and was sweating considerably less than the rest of us. One of the old people from our side had kicked him under the table, and it prompted him to perform.

"You don't have to personally go on Twitter if you don't want to, sir," Winston assured our hobbyless friend. He had taken on the distinct tone of a young person who was willing to sell out. "Many CEOs choose not to. Instead, you can just put your company on Twitter and let it comment on the culture at large."

With those few words Winston had saved the meeting. That was the way to take it, for sure! The *company* could tweet, not any of its people! It would tweet about books and movies and culture! Holidays and babies! Super Bowls, Oscars, and Emmys! We all said a lot of nice things about the progress we'd made over the course of our discussion, and it was agreed that Winston was a real "visionary strategic lead" on the team and I was more of a "detail manager/trainee." After the meeting our pal Winston, newly anointed sellout, made a point to shake each client's hand before they left for the evening, offering up an ass-kisser's heaping ration of enthusiastic-young-person eye contact to anyone who would take it. Winston was riding pretty high on his success until one of the other Patti Hearsts did a quick back-of-the-napkin calculation and estimated that although he would profit approximately $45 for having sat in the meeting and saved it, the old people would each walk away with around a hundred grand in commission. Winston headed directly for the IT closet and began to drink heavily.

The clients promised to follow up with us quickly with a decision, which in client-speak meant they'd follow up with us whenever the hell they felt like it. About a week later some finance guy from business cave in Ohio got on the phone with us and calmly declared that after running the numbers, we would have to reduce our fees by 35 percent to compete with the other bidding social media agencies, but otherwise the contract was ours to win. This news did not go over well with the old people. They responded by staffing the team with only junior people in order to maintain profitability. Commissions again relatively intact, the elders were relieved. Winston, now a seasoned day drinker, was officially put in charge of servicing the contract.

The Patti Hearsts were now left with the task of planning and hosting a "social media kickoff workshop" with all the clients flying back to New York City the following Monday. In a move that we would soon come to regret, Winston encouraged our new clients to "get out there" and start using Twitter for themselves, creating their own handles and experimenting with the service's many features. For the first few minutes things had seemed okay, and our clients even seemed a little excited. "I just uploaded my profile picture!" exclaimed a baking soda toothpaste regional product manager from Fort Lauderdale. "I tweeted at someone!" an associate retail director announced gleefully, having not actually truly tweeted at anyone because "@my friends" was not technically a supported concept. With studious faces, grownups in khakis and no-iron shirts followed a comedian or two that Twitter suggested, such as Carrot Top, along with several news organizations and TED Talk–anointed scientists for good measure. We were, in the words of one of our clients, "Twittering," and everyone was pretty happy. For a moment.

I understood the whole commotion over Twitter in its earliest days myself: after all, Twitter is a venue for impressing people with your smarts, whereas the rest of the world is a venue for impressing people with your looks. The most abusive Twitter stint in my own life had occurred when I fell in "love" with a jobless socialite named Sascha at my friend's annual Fourth of July barbecue. Sascha came from a rich East Coast family and had the kind of physique that anyone could maintain if permitting themselves to eat only lentils, compliments, and grapes. She was never in one place for more than two seconds emotionally or physically, and everyone at the party was watching her with lust, envy, or, in my case, both.

"Please don't tell me you like Sascha!" my friend Michael cried, the host of the party. I hadn't admitted to liking Sascha out loud, but Michael was wearing an Elizabeth Taylor caftan and wraparound sunglasses—he knew exactly what was up. Michael's upset was interrupted by the liminal experience of Sascha bounding over to the two of us while wearing nothing but a scrap of a string bikini. I smiled. Michael stormed away, caftan floating elegantly behind him.

The following Monday morning I resorted to using the only potential advantage I had for winning Sascha's permanent affections: my Twitter account. At my nagging request, Sascha had followed me on Twitter during our fifteen-minute conversation at Michael's party, and of course I'd followed her. I now imagined that she was awaiting my very first tweet the Monday after our meeting. Filled with hope, I pricked my ears to the world and opened up the app on my BlackBerry, hoping to metamorphose my existence into a series of irresistibly fetching bon mots. With downtown New York City as my inspirational cork board and a BlackBerry as my canvas, I obsessively formed 140-character witticisms, chronicling the happenings of my day:

"Is that velocity or speed?"-Anorexic mom using the subway as a teaching tool for her beleaguered 6 year old. G'luck, kid! We don't know!

Who curates Anna Wintour's accent? It is simply fabulous.

"I have a ton of clothes I don't wear but I can't give them to my maid, I have to give them to someone stylish."-a Soho House philosopher.

"Gondola, is that the same as a gazebo? Every time I learn how to do something new I forget something I used to know"-Record producer.

Things were wonderful until I realized I was talking to an audience of no one: the next week Michael told me that Sascha had stopped using her Twitter account weeks before I'd set about impressing her.

Immediately I gave up.

The Tuesday after our clients flew safely back to Ohio we set to the work of announcing to the world that another global conglomerate was taking its parking place at the digital Woodstock of Twitter. Among our clients' requirements: we had to gracefully work in the company's logo, colors, font, and official tagline into their Twitter profile page. As far as the background image options went, our clients sent over a few marketing-team-approved high-resolution photos of their real employees at work. Not one of those photos featured a person of color. Among the "best" options were a photo of two men wearing protective glasses and standing in front of an oil rig and one of a man working with a woman inside a skyscraper late at night trying to figure out the answer to a complicated math problem. In the math-problem picture it appears that the man is correcting the woman's thinking. He seems to be crossing out her work. She is smiling while he does this.

The women on our team refused to allow the second picture any further oxygenation in the world, so it was the people in front of the oil rig who were selected to be the face of the brand on Twitter. We did our best to blur the inimitable towers in the background, sharpening and saturating the employees' faces until their teeth looked like 1800s tombstones if you stared at them for too long. We registered @JumboCorporation as an official, verified Twitter user. To ease Jumbo Corporation into the Twitter world, we purchased small amounts of "high-quality REAL" followers, and in the time-honored, mathematically impossible, unspoken rule that dictates that every Twitter user must have more followers than people they follow, we baited people into following us by following them and then unfollowing them once they followed back. (The tedium of reading that sentence accurately reflects the tedium of building a global conglomerate's Twitter following.)

"What kind of tweet should we kick off with?" Winston asked our group now that our account had dozens of followers and a presumed audience to entertain.

"Something about, uh, the Olympics?"

"Those are next year, stupid."

"Something about being glad to be on Twitter? Like 'Hey, great to be here, looks like another sunny day in the Twittersphere.'"

"We absolutely cannot look that desperate."

"How about something about the history of company. Didn't Jumbo Corporation invent one of the compounds in glue in 1867 or something like that? What did they tell us in the on-boarding meeting again?"

"Pretty sure the only things that Jumbo Corporation invented were created, one, during wartime and, two, for the express purpose of killing people."

"Oh come on, you can't kill people with glue."

"I am absolutely sure you can kill people with glue."

"Prove it!"

"If anyone can find a way to kill people with glue, Jumbo Corporation could do it. I mean, seriously. Who smiles at an oil rig? Isn't an oil rig the place where you go to weep for the fate of humanity?"

"Well, guys, it seems like the subject matter of our first tweet might not be a point of debate anymore because the clients just sent over a press release they want us to promote. And a suggested tweet. Check it out."

An hour later we had written, edited, and approved the following words through the entire client marketing team: "Jumbo Corporation Patents Materials to Present Advances in Spherical Liquid Suppression™ Technology in Copenhagen: COPENHAGEN, Denmark."

And with that, we had our first tweet. It was quickly "liked" three times—not by real people, of course, but by the fake Twitter handles we had set up to unflinchingly support our most favorite client's announcements. The system was now fully up and working.

Each week we set a schedule of discussing, drafting, approving, and then releasing tweets unto the universe on behalf of our client. We had agreed on a formula for determining how many of those tweets should be "lifestyle/fun/seasonal" related and how many should be "topical/category" or "news/marketing material." We scattered the tweets accordingly so our hundreds of followers never knew what they were going to get: An innovation update? A reaction to a sports team? Information about the scintillating Hybrid-Cellulose Change Technologies™ we were patenting up at HQ? Come follow us on Twitter and find out for yourself!

Aside from tweeting out the tweets, another crucial part of pretending to be a company on Twitter was measuring, using social media metrics, how all those tweets performed. We didn't have to compile the metrics ourselves—we outsourced that work. A number of startup companies had already built Twitter "sentiment analysis" tools, and these were impressive in both how expensive they were to use and how consistently wrong the information they provided was. "Everybody loved that last tweet! It got a lot of positive responses!" the sentiment analysis team Data.io or wherever would get on the horn to tell us, feeling great about modern life. Meanwhile, let a few human beings spend an hour researching that same tweet, and they would quickly reveal that Data.io's "sentiment analysis software" did not at all understand sarcasm. And thus did not understand the Internet. One of the tweets our advanced-expert-data-engineer partners were *so* excited about? "JUMBO CORPORATION IS SO GREAT, CONGRATS ON THE OIL SPILLZ. #SLICKMOVE." For a minute we wanted to make an issue out of the crappy "sentiment analysis" software, but we realized that raising said issue would mean also telling our clients that 95 percent of our daily interactions on Twitter were just people making fun of us. That was a problem we did *not* have time for. Instead, we made friends with the sentiment analysis guys. We looked forward to their weekly, factually opposite calls. Our two companies were an ecosystem, farting clouds of crap into the air that grew far stronger when combined.

It wasn't just the one-off, mocking tweets that filled our days; there is a whole organized subset of Twitter that specifically enjoys taking the piss out of the companies who are tweeting—the conglomerates hanging out with a beer and a Hawaiian shirt, attempting desperately to jive with the human youth. Some of this group of troublemakers, occasionally called "Weird Twitter," are quite well organized for scale and can direct hundreds of other people to tweet stupid things in any which way. Our first run-in with Weird Twitter was a memorable one. Per our clients' suggestion, one Monday morning we tweeted the following words: "#Conversation—What are some challenges that #innovation can solve before 2020? #BrightFuture #JumboCorporation #Ideas." Dozens of people quickly responded in agreement when one person postulated that if grizzly bears could somehow be forced to evolve at a quicker rate—using our scientific research and medicines, of course—then they could perhaps be trained to ride bicycles and form their own nonprofit companies, creating a low-carbon-emission taxi system for the people of the future. Dozens of other Twitter users added to this dialogue with factoids, charts, and Photoshopped "images of the future." Da Vinci's bicycle was quickly re-engineered to fit the physical requirements of the average bear. Other users quickly formed Twitter accounts for famous scientists, most long dead, and posted, pretending to agree with this general line of thinking. At one point a scroll of promising, fake grizzly bear research emerged out of the Bermuda Triangle, brought to our attention by the extremely dead but suddenly tweeting Italian explorer John Cabot. Weird Twitter was working in full force.

Our relationships with these hecklers were always interesting because although they were publicly making fun of us, they were also creating heaps of metrics that we could report back to our client, thereby yielding "success." Most often we goaded further responses rather than breaking the party up with a "Hey, let's be SERIOUS, guys" type of rebuttal. We could laugh along! We were also the same age and pay grade as our heckler counterparts. Like them, we gave no fucks.

"Thanks to your wonderful 'innovation' topic idea," Winston reported back to our clients later that week, "This week we have been enjoying a lively back-and-forth about the future of innovation at Jumbo Corp. Several subject-matter experts—including one scientist—have chimed in from across the global Twitter landscape." Ever terrible, the social media monitoring software team over at Data.io was indeed charting a huge positive uptick in daily interaction rates along with the presence of many positive words, and *many* positive hashtags. The clients were extremely pleased. We were all getting the hang of this Twittering thing, weren't we! The clients also appreciated Winston's attribution of the idea back to them. Life was as good as it got on a combined client and social media agency team.

With Winston at the lead, the rest of us did our best to assist with the deception and contain the laughter throughout the duration of the call. Later that day, when my friend Courtney called me to finalize our plans for the evening and she asked me about how work was going, I finally popped. "Well, we are currently conversing with several Los Angeles–area comedians about whether grizzly bears could someday contribute to the global economy."

"Oh, wow, I see, that's so interesting," she said, playing along, fully aware of the stupidity that is whatever any young white-collar New Yorker is doing at his or her job at any given moment. "And as you know, Courtney, many of these comedians view trolling America's tweeting corporations as a nascent form of performance art," I continued. "Which is why there currently is a petition circulating online that proposes we give six grizzly bears iPhones and have Jumbo Corporation study the results. Feel free to e-sign it! You can locate parts of the lengthy discussion under the hashtags #bearsinnovate, #bear-conomy and #JumboBears2020."

Later on, it would turn out that the innovation-yielding, bicycle-riding grizzly bears were our best social media statistics of the entire year.

The word that really generated all the Fortune 500 excitement over Twitter is that nasty word I have mentioned so many times before: innovation. Innovation. *Innovation.* #Innovation. Innovation is a horrible word. If it weren't for Innovation, I am convinced that Twitter would not be publicly traded on the stock exchange today. You see, using Twitter requires some level of truth telling and transparency, and because that is 100 percent exactly the opposite of what many corporations want to do, what they really needed was a way to be transparent and also constantly lie by omission at the same time. Somewhere in Brooklyn a marketer in a meeting discovered that although not every company has ethics, *every* company has innovation. If you make cigarettes, nuclear bombs, or replacement components of illegally traded AK-47s, well, then, you are contributing to #innovation. A company needs computers to do all that stuff, and anything with a computer is #innovation. Problem solved! Countless conglomerates have delightedly rebranded themselves. Some of these companies really do build interesting shit, sure. But most of them don't. Virtually all of them tweet about it, though. There isn't a massive conglomeration in the world that hasn't made a home for itself on Twitter, prattling on and on about all its terrific innovation.

The biggest problem is that no one holds Twitter accountable for anything. No CMO or CEO ever calls up Twitter and says, "Hey Twitter, I looked into your practices a little bit, and I've discovered that by allowing for the creation of *millions upon millions* of fake Twitter users, you're constantly inflating your own metrics." It's not as if any Fortune 500 executive would have to look far to start asking these questions: after Twitter went public Nick Bilton wrote an excellent piece in the *New York Times* explaining how easy it is to create fake, "active" Twitter users that are actually nothing more than software. Mr. Bilton went so far as to use the words "giant pyramid scheme" in describing how much money is made by creating armies of fake users to do corporate bidding on Twitter. To see the level of scrutiny that Wall Street and the Fortune 500 devote to Twitter (absolutely none) is to be unsurprised when company-wide scandals like Theranos appear. Technology makes corporate America fearful of asking the simple questions; no one wants to look dumb in front of a kid wearing a hoodie. As a result, corporate America sometimes gets exactly what it pays for.

There were always "oopsies" moments for any big corporation attempting to delicately expose the more attractive parts of its hairy underside on

Twitter, and our account was no exception. It is hard to make ugly things seem pretty and friendly, but add in that those companies also have to seem ethical, environmentally responsible, and civic minded too, and you have yourself a true shitshow. Thankfully our social media agency's old people had forewarned the clients that these sorts of moments might happen. The clients were told that we would "fail fast" and learn from our mistakes. Our biggest oopsies moment happened on the anniversary of the USS *Maine* sinking in Havana Harbor. Why we were tweeting about this topic, I cannot remember, but we were. "It's a sad day for #reflection. Let's all take a moment to #RememberTheMaine," we'd tweeted alongside a colorful rendering of our logo anthropomorphized and carrying an American flag, designed expressly for this particular occasion. Whether the clients had fully approved this tweet became an issue of hot debate over the following hours because the Twitter kids had an absolute field day making fun of our drunk Miss Hannigan of a tweet, garishly adorned in hashtags. Unlike the time we innovated bears, the media picked up this tweet, and its failure therefore became unavoidable topic of discussion in our client update. Either we told them or their miserable teenage kid did later that night at dinner, so it had to be us. One full day later the Mainegate joke was still marching proudly on. A ruthless company was taking a pause to think about something that happened more than a hundred years ago, and we'd handed the jokes directly to the Twitter comedians. People diligently Photoshopped our flag-wielding logo into all *sorts* of important moments in history. Naturally Normandy came first, but I'm sure you can imagine what some of those other special days were. To this day this mistake still gets mentions in articles written by social media pundits about how *not* to use Twitter.

So why do idiotic tweets about things like the USS *Maine* sinking ever even get made, let alone approved? The truth is that it's really hard to be a talking company on the Internet seven days a week. You run out of things to say somewhere around week two. You become willing to write about anything you think your client team will approve. That's why all social media agencies desperately scan calendars looking for any subject they could talk about in order to avoid the activists, casual protestors, and general big-company-haters that make up many of Twitter's users. Nonreligious holidays are the finest sorts of tweet-worthy distractions, as are corporate-created holidays that encourage the rabid, meaningless consumption of goods, like National Sibling Day, National Donut Day, and Black Friday. "#BlackFriday will be here before you know it!!!" the brands of Twitter all elbow each other to tell you first. Coupons, offers, and influencer promotions will follow.

The worst thing a brand can do on Twitter—when faced with the hell of the blank tweet box that always needs filling—is believe that the medium is an outlet for young employees' creativity and let them use Twitter only for corporate-branded jokes.

Taco Bell is the reason why corporate in-joke tweets exists. Taco Bell is the "funny," "cool," "Wow, they really GET Twitter" brand that is most frequently cited as "WINNING" on Twitter. Taco Bell is hilarious. "Is butter a carb?" Taco Bell will tweet, digitally adapting some all-meme-knowing, tween-girl-with-an-ironic-Swatch-watch hellscape of a persona in its 140-character constraints. "I speak four languages: tacos, Fire Sauce, burritos, and Baja Blast." "Dating is cool, but have you heard of @TacoBell?" "I wish I was full of tacos instead of emotions."

Taco Bell takes the notion of corporate personhood to planets so depressing, even Christopher Lasch couldn't foresee them: sometimes the brand likes to ask questions to other brands, as @TacoBell did in this moment when it asked "@OldSpice Is your deodorant made with really old spices?"

For all our mistakes—and we made many—we did not try to "show off our youthful humor" via the character we created for our client on

Twitter. We didn't engage in stupid conversations with other brands. We were just trying to play the role of a benevolent tweeting conglomerate. At least we behaved ourselves.

After about seven months into the round-the-clock tweeting as the digital persona of Jumbo Corporation, our account had amassed over fifty thousand followers. Even at that size, most tweets only got three or four likes, and it was rare to get multiple retweets. Our clients had become wise to these analytics and set a call with us to discuss how they could become more "involved" with the Twittersphere. Because we weren't able to garner any attention by sharing our truths with the world, we were going to have to buy it. In careful language we explained that "collaborating with influencers" often involved "some form of remuneration." The clients agreed and set a budget of $50,000 for purchasing influencers' tweets.

The next day we made a list of Twitter "influencers," meaning Twitter users who had amassed more than one hundred thousand followers. We doled out contacting them and pleading with them to please engage with our stupid social media campaign. We emailed them, Skyped them, called them, used any public contact information available to make our case. The conversations for all our conglomerate's myriad products went something like this:

"Hi, I'm working on a marketing campaign and am wondering if you could find something positive to say about Calclear eyedrops to your 1.2MM followers? Calclear Drops are one of the many reliable products made by Jumbo Corporation, and we're trying to get the word out with cool people like you on Twitter. I've noticed that you wear glasses—do you use Calclear Drops?"

"Go kill yourself."

"Would you want some free products in return for helping us out?"

"Um, I only work with brands that I like."

"What if we invited you to a special influencer dinner full of other young influencers here in New York? Would that sweeten the deal?"

"I meet influential people every day," the empowered makeup-tutorial-creating tween on the other end of the phone line would growl back at us, well trained by her wise mentors, the Kardashians, to rebuke any low-ball endorsement offers and hold out for the best—the detox teas, the teeth whiteners, the waist trimmers—the companies with the real bucks.

Let me get this dynamic straight, we would think to ourselves after the twenty-seventh downer of a call, *you get paid to tweet stuff about brands, and I get paid to get you to tweet stuff about brands,* we Patti Hearsts would think to ourselves on the other end of that phone call. *We're all funding our nightly wine and burritos because Grampa Corporate America doesn't realize that "Twitter" is a bunch of malarky.* My most embarrassed moments always occurred when I was on the phone with someone whose work and wit I admired, someone who I'd choose to follow on my own Twitter account. My fan-ship was publicly available data to the person on the other end of the line, the one who was spiritedly rejecting me. I always tried to believe in the power of my own rant—that we were the same—but often, as I listened to the delight in the voice of whichever social-media-influence-wielding young person was on the other end of the phone line as he or she rejected me, it felt untrue. After repeated failures with Operation Influencer Outreach, we reverted to buying more fake followers to bolster the lag in momentum and building more fake accounts that could "like" and "retweet" all things we tweeted. It was easier that way.

Nine months into the project the motivation to continue on was drained, and we sensed an imminent end to our project. "If a tweet's tweeted by someone who isn't themselves and then that tweet is favorited by someone else who also isn't themselves, is it actually a tweet?"

one of the Pattis posited later that afternoon as one of the robots we'd purchased retweeted one of the things we'd forced @JumboCorporation to say. The kid asking this question had graduated from an Ivy League.

"Yes," said Winston evenly, playing his role of the dad to our group. "It is our job to say yes. If it looks like a tweet, it's a tweet."

"You know my mom emailed me asking for help 'making her tweet better' the other day?" Winston told us over our four thousandth round of drinks, now finally drained of his desire to emerge as a promising young leader at our social media agency. "In my spare time I am officially helping to build my mom's social media brand. Aside from the fact that she gave me life, she has become no different from our client."

"It's not that bad," one of the graphic designer kids replied. "She's a human being who's trying to socialize on a stupid new service. She's not on Twitter standing blindly in front of an oil rig, as if she has no idea what's behind her."

In the end of my stint as a social media executive, in the last days before I would be able to find work as a person who wrote about technology rather than a person who wrote inside of one of its stupidest systems, it was the older folks at our social media agency who saved us. Once the clients had started to feel that they were paying a lot of money for something that wasn't really creating a noticeable result, we waited for someone to correctly blame the failure on us. But our agency's executives stepped in to absorb the blow, finding a business way of breaking up with our client but still sort of staying friends (aka never speaking to each other again). The guy who spoke at the speed of acronyms did his job—floral shirts, graph paper, and all. Some of the young people quit when we lost the account; one was laid off. Twitter spun madly on.

What is Twitter? Twitter is a place to shout. Twitter is where a brand goes to humanize itself online while paying its workers obscenely low wages offline. Twitter is a place to pretend-talk, to favorite, and to #innovate.

@JumboCorporation still exists (well, the handle this story is actually about does, anyway), and it's still active. It's weird and, I guess, nice to have a Google-able testimony to our stupid project's existence and that particular time in our lives. I won't name the handle, and I have blurred details specifically to avoid any upset, but the next time you see a company tweeting away, take a minute to think about how that tweet got drafted, approved, edited, and posted. Picture all the twenty-somethings in their poorly put together business outfits from Zara and Forever21, using their college degrees to get lists of hashtags approved. Think about all the people sitting in the conference rooms of America, and try to imagine how little impact their collective effort will have on a service that's mainly a lonely crowd of robots. And then call your accountant—to make dead certain that you don't own have any financial stake in #TWTR.

The People I Have to Stalk on Facebook

(Or I Can't Get Anything Done)

OW did anyone concoct the real idea for Facebook? It is a brand-sponsored version of the US Census that politely knocks on everyone's door and says "Hello! In order for people from high school to stalk you, could you please tell us the following information: *Are you fat? Single? Employed? A deadbeat parent? Capable of travel? Sexy?* We ask these questions because you have one hundred to two hundred former classmates who would like to not pay to have the privilege of sometimes stalking you when your name comes up in conversation. Oh! In exchange for your information, they are willing to let you stalk them too."

When I have writer's block one surefire way to make it worse is to log onto Facebook and see how some of my favorite personalities are doing. Facebook knows I search for their names and never comment on their photos, praying to leave no trace of my time on their page by means of accidentally hitting the "like" button. (Facebook: you should have an official setting for this, sort of like Clippy the Paperclip from the old Microsoft Word days. *Hello! It looks like you're stalking some-one. Want us to make sure you don't accidentally "like" their life?*)

In no particular order, I present my most important people I stalk on Facebook in order to not get anything done and then hate myself so much that eventually I do get something done that day, even if it's only paying my American Express bill.

Perfect-Life Couple

My friend Joe admires a specific bracket of attractive people that he classifies as being "the level of pretty where everything in life just happens for you." Not a model, but a gorgeous person with a normal job; the type of person who will innocently chalk their success up to hard work—*not* their hypnotic facial symmetry—and truly believe him or herself to be one of the normals. I met one of these people at a party once, and now I stalk him on Facebook.

Thanks to Facebook, I've learned that my Perfect-Life Person has become half of a Perfect-Life Couple: his girlfriend is thin but quirky looking (which makes him more perfect, as he is capable of loving someone not as perfect as he). She's a yoga teacher, of course, and on weekends they make their own soup. In their earlier years as a couple, back when I first started stalking them, they regularly dreamed aloud of someday getting married and moving to Hawaii. Because they are the Perfect-Life Couple, I can report that they have achieved those particular aspirations and more. I seriously cannot take my eyes away from this couple. And thanks to Facebook, I don't have to. They have no idea who I am.

Weren't You a . . .

Remember those girls you and everybody ignored in high school? Of course you don't. I didn't remember them either until one of mine moved to New York City and became an incredibly high-end stripper. Or something like that. It's really not polite to ask. And thanks to

Facebook, I don't have to. I can just scroll idly and wonder why she's been in the Hamptons for four straight weeks even though she doesn't have a job. Not to mention the fact that she just spent half of last month on a yacht off the coast of Amalfi. Like anyone who maintains a Paleo-lifestyle-themed Instagram account, this young woman is in enviously good shape. She wears a bikini to wash her car, and while the rest of us toil away at our day jobs, she enjoys buying seasonally themed outfits for her Corgi and then posting pictures of him wearing them online. Her Corgi's name is Honey Molasses, and he has thirteen thousand followers on Instagram. As a deeply nervous person, I worship my former classmate for not giving a single solitary shit about anything I care about—maintaining a steady job and income, the probably nonexistent expectations of friends and family, taxes. Her infrequently updated fitness-lifestyle-travel blog includes a permanent disclaimer on the bottom that reads, "If your going to email me on grammar, just don;t. Your a loser and people don't like hanging out with you, seriously."

Honestly? I *do* email people and about their grammatical choices, and I *am* a loser. She makes a completely valid point.

People Who Dropped Out of Society and Are Hiding Something

I have a few of these people in my Facebook feed. They're hard to spot at first, but once you're in, you're good and in. One of mine is a former high school popular girl who achieved her biggest lifelong dream of becoming an assistant makeup artist on a local CBS news affiliate, and then she totally ghosted. What happened? Why was she last seen on a mountaintop, prepping a weatherman to update America on the wind conditions at the Special Olympics? What happened next? Why won't she post something? She looked perfectly normal in that last picture, which I know because I've searched it sixteen times.

The most mortifying person I ever dated also fits into this category. *Relatedly . . .*

The Deeply Religious

Jesus and Facebook go together like chicken and waffles. I truly love to see His big ideas memorialized as social media status updates, especially when hastily applied to a relatively mundane situation in modern life. One example:

"For he who has walked aside me hath not seen their own footprints, for in I was carrying him in my arms."-John 41:41 Wish me luck on my midterms guys!!!!! #feelingblessed #HeIsEverywhere

I am comforted by the presence of the Deeply Religious on Facebook. First and foremost, I enjoy their inherent hypocrisy in that they feel superior to those of us in the nonreligious world but simultaneously crave the warm heat of our attention, via Facebook, just like a regular old glutton.

Also, sometimes the New York media sarcasm that I read all day long becomes too much to bear, and I just want to read something written by someone who isn't in on their own joke. The online religious are certainly good for that.

People from My High School Drama Club Who Are Still Pursuing Careers in Show Business Even Though We Are All Thirty-Five F*@#ing Years Old

Imagine getting married and also having to go on a first date every single night for the rest of your life: that is what I imagine it is like to be a thirty-something unfamous actor who is living in Los Angeles or New York.

I have mixed emotions about the people from my high school class who opted to pursue the dramatic arts. There's the noble side of my emotions (I'm impressed by anyone attempting to stay true to their dream for more than seven minutes, especially when that dream involves staying constantly skinny despite not making enough money to employ a personal trainer). But let's ignore the nice emotions and instead focus on my more horrible emotions. Like about how goddamnned tired I am of the drama kids and their plight.

Drama kids who were born into the 1990s were somehow misled into believing that the world wanted to experience their creativity. During our high school years these kids discussed plays with a sense of authority that was totally disproportionate for someone whose sole theater-going experience was seeing a matinee of *Rent* one time with their parents. At my high school these kids went into Hartford to take private acting classes on weekends. They "invested" in their "career" during their teen years, which in reality means they "got head shots taken once." Some drama kids were so self-righteous that even to this day, fifteen years after high school, I slightly delight in reading their complaints about how hard it is for them to pursue acting. Shakespeare may have said that "all the world's a stage," but he would not still mean those words, oh no, not if he could have lived to see the selfie stick. Shakespeare and his contemporaries endured surgery without anesthesia. He would not allow perfectly capable people using Facebook to complain about the fact that it's really hard to land a TV pilot if you don't know anyone on the inside. Oh no. He would not have time for that.

Hateful Former Coworkers

It's always nice to watch people you don't like as they age. This is when Father Time really is on your side. He is creating comedy! It's sarcastic! Grab a glass of wine and log onto Facebook for a front-row seat.

Remember the IT guy who used the server room to roll his own cigars? He's on there, and if you've got an hour or two, why not check out what his last six vacations were like?

How about the special lady with the pixie cut who considered herself to be the office's unofficial grammarian because she'd once self-published a book of erotic poetry? She's on there too. Or the boss who was so impressively disengaged from the process of raising her kids that, she made the nanny take them to Disney World? There's always time for that.

Relatedly . . .

Fantasy Football Fatsos

Does your hometown have its own *Friday Night Lights*? In other words: Does your hometown have weathered and beaten adults who walk around talking about how great life was for the five minutes when they were young, in shape, and athletic? Back in Connecticut we've got these guys in spades. Shoot me an email, and I'm happy to provide you with some delightful accounts of folks who, when they were athletic-jacket-wearing youngsters in the nineties, heaped insults upon our town's biggest nerd, who is now a pediatric oncologist. But who's keeping score?! These men have now taken their rightful place in society, which is peddling affordably priced cubic zirconia engagement rings, assessing the damage incurred during minor car accidents, and managing the night shift of Subway sandwiches. And they are on Facebook. And they seriously don't seem to understand what happened.

I Had a Baby and Now I'm Born-Agains

Also known as all the girls from high school who now wholly define themselves by being a mom. It's really weird to watch the people you remember as being uninteresting classmates now obsessing over being moms.

These were the girls who just didn't care. They smoked at precocious ages, they sat in the back of the classroom, and they habitually avoided any sort of learning. And now they're MOMS. Sometimes it's all too much to comprehend.

Anyone Redecorating Their House

I cannot turn away from a person who is publicly redecorating their home. Forget the weight-loss documentation photo, the weirdly intimate sexy wedding night photo, or the bare pregnant belly photo with a flower painted onto it—all of those types of overly revealing photos are *less* overly revealing than the Person Redecorating Their Home. And oh, it is so much fun to judge! (Or to experience the hot-faced intensity of envy of a home that is better than yours.)

Several years ago Pinterest taught the world that anything can be organized in a thousand mason jars, that adults should use yarn, and that no matter the purpose of a room, covering one of its walls in corkboard material is a good idea. And Pinterest didn't stop there! A spray-painted cheese grater is a whimsical organizer for dangly earrings. An antique watering can? Just waiting to become shower head. With ideas like Pinterest's circulating around the universe, who knows what surprises a redecoration can bring. (PS: Pinterest is one of the "most valuable" startups ever created. Yes, we are in a bubble.)

In this modern, confusing world, the simple binary of being able to judge someone who is redecorating as either having a much better or a much worse life than you is a true comfort.

Anyone Who Hashtags Their Baby

Why would anyone give a hashtag to a baby? I don't know. But a certain breed of thirty-something, just-digitally-savvy-enough parents on Facebook and its popular-cousin-who-does-drugs Instagram are indeed adding hashtags to the bottoms of their pictures of their babies. Couples who hashtag their babies are almost as bad as couples who create separate social media accounts for their babies immediately upon birth. Both groups seem to harbor a belief that their genetic creation will soon be of such import to the world that the traceability provided by a hashtag and/or devoted social media account will be crucial gestures to future generations.

I've noticed that couples who assign their baby a hashtag are usually the sort of people who allowed their dog to participate in their wedding as a ring bearer or who photograph themselves while exercising. I've never given birth, but I'd like to think that if I did, I would not take my newborn, barely human blob, strap a bow on its head, direct its attention toward a camera for a photograph with some light-to-moderate finger snapping, and then add a hashtag on the bottom of the whole horrible ordeal.

Relatedly . . .

Couples Who Hashtag Their Vacations

You're going to Africa for A WEEK. Maybe two. Either way, you do not need to rebrand the fucking continent. If you want to let everyone know you're in Zanzibar, use the provided geolocation features. Do not also add a second or, god forbid, third hashtag like #TheGordonsDoZanzibar, #GordonsWanderlust or #Gordons#3of10BucketList. We get it, okay? WE GET IT.

People Who Teach Their Kids to Do Yoga

This one is just insane to me. I can't look away. Yoga is what we learn to do when we finally get old enough to truly fear having a heart attack. *Goddamnit*, we say to ourselves, *I have to learn to calm the fuck down so I don't have a heart attack. I guess I have to go to a yoga class.*

What world of stress does a kid need to escape from by means of learning to do yoga? And holy shit, is your daughter *meditating* right now? Because my kid seriously has not stopped asking questions since 2010. He asks questions in his sleep. I am not even kidding—he will ask questions to whatever adults are in the room as he *sleepwalks*. That is his level of relentlessness in life. Meanwhile not only are you managing to stay in shape as a parent by doing yoga (something that is already so amazing I could never take my eyes away from your Facebook documentation of your abs right now) but you're also teaching your child to do it alongside you too. Because you care that much. I too have all the time in the world for this kind of stuff. All the time in the world.[4]

People Who Are Constantly Traveling Around the World But Also Not Explaining Some Crucial Component of the System of Their Life to Everyone

Above we discussed the fascinating Former Invisible Classmate Who Is Now A High-End Stripper component of this delegation, but there are many more members. Oh yes. I have no fewer than four friends on Facebook who've made it their life's nonwork to constantly travel around the world and post photos about it. However, as seemingly transparent as these friends are about their physical locations, other massive components of their life go entirely undiscussed. Who paid for

---

4. *I do not have time to teach my kid yoga.*

you to fly from Sanibel Island to Croatia in the last forty-eight hours, enabling you to take your top off in front of that waterfall and post a picture of your naked backside, embracing life? Is that person your boss or your boyfriend? Are you really the champagne-drenched guest of Carnival Cruise VIP royalty that you make yourself out to be in those photos, or do you actually work on that boat? Is that really your cute beach house that you bought while not having a job, or is that where your mom lives and also she sends you money? Facebook, can't you find some way to get us these answers, please?

The Prom King on Instagram, Ten Years After the Party

WHEN I was a junior in high school I decided that I wanted to become popular. Fortuitously, my scientist parents were about to make the one wanton decision of their lives: I would be left at home unsupervised for a weekend along with my little brother.

Normally ones to frown upon any vacation not spent in a tent, Maine was an exception to my parents' rules because it was a place that turned holidays into work. The two of them had planned a quintessential rise-at-dawn, sleep-by-dusk Maine experience: stilted breakfast conversations with strangers at inns, jaunts to folksy outlet stores, and long walks along punishing rocky coasts. Amid all the excitement, it never occurred to either my mother or father that leaving an ungrateful teenager home alone with a car, a finished basement, and over $3,000 in personal savings was a formula for total disaster.

I was a barista at my town's only coffee shop, hence my fortune. Every weekend from 7 a.m. to 1 p.m. I served drinks and snacks alongside Leila Rodriguez, the most popular girl in our entire high school. To my complete surprise, Leila and I had become some version of friends. It was our new manager who'd first bonded us together: Mean Carl.

Not understanding Leila's status as teenage royalty, Mean Carl had sternly admonished her one Saturday for not saving the leftover cooked bacon: in his eyes the meat could be resold the following morning. I responded to the incident by stealthily carving my initials into every crisp strip of pig I could find, which I informed Leila was "research for my phone call to the Department of Health." Leila suddenly thought I was hilarious and invited me to her house for brunch. I toured her bedroom and hung out with her stepmother, who wore a track suit and drank glamorous pink wine to mark the passing of the morning. "It's five o'clock somewhere!" Mrs. Rodriguez said to me, and I laughed wildly, like someone who'd just seen a monkey in a business suit fall down an entire flight of stairs.

Once the vacation weekend arrived and my parents were safely puttering toward L.L. Bean country, I headed to the coffee shop and told Leila about my exciting, supervision-free living situation. It took approximately not even one minute for her to suggest that the two of us host a party together on Saturday night.

"We can make Jungle Juice!" Leila squealed, clapping. I didn't yet understand that the popular kids had elaborate underground systems for procuring huge amounts of the beer and liquor that they considered to be human gasoline. Leila immediately began calling a stream of high school seniors on her cell phone—a feat that was absolutely amazing in 1999, when there were members of the CIA who did not yet have cell phones. Leila concluded that we would need $350 to buy the proper amount of ingredients for our Jungle Juice. "Justin said he'll get everything for us—Justin is sooooo nice," she told me. Justin was the collegiate older brother of our class prom king. As a thousand-aire, I saw $350 as a trivial amount of money, so I ran down the street to the nearest ATM. Jungle Juice. By name it sounded like a collection of truly exotic ingredients, probably fruit juices indigenous to Hawaii, rare sweet alcohols, maybe little umbrellas floating in each glass. I was proud I could afford such tropical extravagance for my classmates.

The next afternoon I picked up my little brother after school in my Mazda Miata (the car that James Dean did not drive in *Rebel Without a Cause*) and at long last revealed to him my evening plans. "Some of the most popular people in the world are coming over to our house tonight," I said, managing to sound both breezy and foreboding. "So I'll need you to sit upstairs by yourself all night long, guarding the phone in case Mom and Dad call."

Once home I raced to my room, turned up some of my favorite Bette Midler music, and began the work of selecting an outfit for the best night of my life. I put on my newest pair of sheer black pantyhose from the drugstore, a black miniskirt, and some clunky black platform shoes. I finished the look off with a green T-shirt because Leila had sternly reminded me it was St. Patrick's Day—an occasion as religious to the well-muscled, white-hatted social leaders of our teenage community as the "Saint" in its name implied. I looked in the mirror and realized that my outfit was exactly what I would have worn for a chorus concert. Determined to rebel, I rolled up the skirt as far as it would go and ran downstairs, where Leila and her army were hauling gallon after gallon of Everclear, a grain alcohol, and Stop 'n Shop store-branded Hi-C concentrate into our basement.

As these exotic Jungle Juice supplies were piling high in my father's office, a woodworker's creative respite turned into a alcohol laboratory. Before the mixture could chill, kids began to make their way up the front lawn like a horde of Abercrombie locusts. The back porch quickly transformed itself into a smoking depot, my brother's bedroom a quiet space for couples to form or break up, and anywhere a great location to leave a mostly empty beverage container. Two hundred of my new-found popular friends were carrying on, revelrous, into the night. They were drinking to escape the hell of a good education, fresh air, and supportive parents who cheered through soccer games, and it was all because of me.

The only problem with my party, I embarrassingly admitted to myself, was its lack of ability to meet even my smallest of expectations: no crushes revealed themselves, no exchanges worthy of John Hughes were had, no one even really said hello, let alone thanked me. Kids used my parents' shit without asking, muddy footprints tracked in all directions, and our pool table became an amateur stripper's practice stage. Eventually I pushed my way into the corner "bar area," hoping to sample some of the gluey, jungle-themed beverage that two months of my part-time employment had funded. At the front of the line I discovered Justin and several of his blockhead compatriots charging money for the alcohol I had paid for. "Five bucks each!" Justin shouted to the crowd in front of him, shaking a tube of red Solo cups. Justin was the type of kid who wore layers of sweatshirts: one foundational sweatshirt and then another bigger sweatshirt on top of it, this one with the sleeves cut off. When he noticed me approaching he began to wildly wave his hands. "Yo! Yo! Yo!" he called out, clearly searching for a substitute for my name. Pointing at my face, he eventually managed, "You get a drink for free!"

Disappointed, I took my cup upstairs to look for Leila. Instead, I walked right into the Prom King, who was standing outside my parents' bedroom. The moment felt as contextually logical as a Dali painting. "What up," he said to me. I smiled, hopeful. His face was stern. "Your little brother says he has to sit by the phone all night and can't come downstairs? That's messed up, man. He's your brother." I walked into my parents' bedroom to see that the Prom King had passed a considerable amount of time with Robbie, playing numerous rounds of Tic Tac Toe.

It was just past ten o'clock when the police showed up. Due to the social nature of my after-school job, I knew both of the uniformed officers by name as well as their coffee preference. Defeated, I took a moment to roll down my miniskirt before the confrontation began in earnest, and then I started to cry.

My sentence was two hundred hours of community service, to be completed over the course of a year. The judge had lobbed me a softie punishment because I was an honor roll student and a first-time offender; my parents, bearing fresh farmers' tans and granite facial expressions, probably seemed fully capable of punishing me enough themselves. To fulfill my civic duty, every Tuesday and Thursday I played piano at the local Drama Club, where middle schoolers sang about love affairs they didn't understand and started fights with each other about who was more talented. Leila eventually started dating Justin in an official capacity, quitting her job so she could spend more time with him.

It was almost a month before Officer Antonio came into the coffee shop for his usual bagel and skim single latte. Sheepishly I muttered some form of apology to him, dreading every flicker of eye contact, but Officer Antonio did not seem to be too upset with me. "You're a good kid!" he said, assuming me innocent, not knowing the moral compromises I'd made to host my soiree or the fact that I still didn't entirely regret hosting it. "In a few months you'll never see any of those lousy drunks again!" Officer Antonio was exactly right—until ten years later when he was exactly wrong. Cue 2010, the year my former classmates joined the new service known as Instagram.

After a decade of relative silence (we were too old to post anything of substance beyond the occasional belch of wedding photos on Facebook), Instagram was a communal rebirth. All of us together again, grown up and uncool, struggling with hashtags and filters and attempts to capture the glory of our sunsets in a single digital frame. The flawless blonde from the sacred lunch table drove a minivan, and the kid who once stole a license-making machine from the DMV was now a Christian youth counselor. Soon we were all using the application constantly, documenting our kids, pets, and home improvements. On the weekends we drank legally procured beer on our decks and admired the springtime blossoms of flowers. We zealously participated in newly formed virtual holidays such as Throwback Thursday. We took extreme close-ups of onions and tomatoes stacked neatly on cutting boards, providing a very liminal sense of how we in particular "make chili," despite the ritual remaining relatively unchanged since the dawn of the spice route.

One thing I know for certain is that we all look dumb as we take ourselves so seriously through Instagram, filtering our ephemera. It is funny, on the simplest level, to see the Prom King take a picture of a salad. It's funny that Justin still layers his sweats and hangs out with his same refrigerator-shaped friends who have come to closely resemble the marlin they so love to fish for on weekends. Leila lives close enough to him that their families sometimes hang out—kids and spouses, all together. Although first we all went wild with the wisecracks, now it's just life. We tap twice to "like," to say "Hey, I thought of you—for a second I acknowledge you exist." The platitudes we were all incapable of exchanging in high school have been usurped by emoji-hearts under a costumed pet.

I was almost a child of the previous generation: those whose high school experiences warped into night-watch plaid memories, forever held in their consciousnesses. Nowadays we're able to realize how silly it was to deify certain members of our social scene: everyone's too busy taking pictures of hamburgers for previous exchanges to hold any power. We said nothing meaningful to each other during high school, and we judged each other to be different. Now, on the Internet, we say nothing meaningful to each other again, and we realize we're all the same.

Match.com Is Like Putting Rosie from the *Jetsons* in Charge of Your Dating Life

DURING my twenties I lived alone on the second floor of a dirt-soaked East Village tenement. The building's entrance hallway was decorated in abandoned phone books and cigarettes, the one-room apartments filled with defeatist coffee tables and pull-out couches left over from the previous occupant. If I had to design an advertisement for the place, I would make a video of the Sunday morning when a cockroach flew straight out of the shower drain—as if my innocent-seeming bathtub was actually an unmarked portal to hell—and then proceeded to zoom around the living room until it died three hours later, stumbling through the nest of dusty cords behind the TV. Mine was the kind of residence where you walked in after a long, fruitless week of work and immediately ordered a case of Trader Joe's purest 99¢ wine just to survive. My "work" was a job that should have been completed by a robot, not a human being with a middle school diploma, and my love life was an intensely one-sided, semi-imaginary relationship with a straight woman who only slept with me when she was drunk. I also had no savings and a chronic case of adult-onset acne.

Yet despite all this, I somehow wanted more from my life. Specifically I wanted to fall in love and avoid spending my sixtieth birthday alone and sculpting a cake out of cat food, as I'd begun to suspect my parents believed was my fate.

My mom and dad raised me in a Connecticut suburb that over the years was just big enough to have produced one notable television star and seven lesbians. When I was eighteen years old I didn't even realize I was a lesbian. To my knowledge I'd only ever met a lesbian once in my life—at a church fair about Being Different. The last thing I ever wanted to be was someone "different" enough that it required standing on an altar explaining the ins and outs of my personal choices before an audience of grim-faced matriarchs in L.L. Bean skorts. My difficulties with love were also created by the fact that I'm also not the easiest match by personality: I've always had isolationist, stubborn tendencies. The photo my mother cherishes most from my childhood was taken in our backyard was I was seven and in the wake of a violent temper tantrum. I am sitting on the roof of my nine-foot-high playhouse wearing a sparkle bicycle helmet, overalls, and a G.I. Joe belt. I am fearlessly looking right into the camera. I am crying. I am also holding onto a book that I might read later while perched on top of my roof, as I'd sensed that any conflict might not resolve quickly and detente would require entertainment. When it came to finding a partner, it was only New York's numerical odds that seemed strong enough to work in my favor. Squint and ignore all the steel, and New York City is just one giant pile of people all stacked up on top of each other.

From my first night as a student at NYU on, I'd set about the business of meeting people, going out for drinks with anyone who suggested it. I wasn't hurting for invitations, as New Yorkers would ask a barn owl out for drinks if it could pony up a MasterCard and pitch in for the second round. Drinking is New York's official pastime; our state flower should be the grape leaf. The problem was that I didn't want to admit aloud that I was a lesbian with zero experience. I didn't want to go to a gay bar out of fear that I'd be immediately identified, like a telltale stain under a cop's blue light. "This one's a faker!" Instead, I headed out on the town in unspecific groups, crossed my fingers, and hoped.

My first fake girlfriend was named Harper, and we met at a loft party in Soho. Harper was sitting on top of a table explaining Marx, whom she'd never read, after having dropped acid and eaten a bouquet of roses. Harper *was* the party. When she announced her intentions to make out with me later, the whole room cheered. While I pounded breath mint after breath mint in eager anticipation of our coupling, Harper devoted the next six hours to consuming something called "SoCo and Lime" in scientifically impressive quantities. "SoCo and Lime" sounds like the whimsical name of a good-times drink, but it is actually the legal equivalent of Oxycontin. By the time Harper was ready to make out with me, she was missing a shoe.

When I wrote a story about the matter, Jack—my favorite creative writing professor at NYU—affectionately deemed Harper my first-ever "Maniac Girlfriend," declaring that I probably had a weakness for the type. His online and offline research left him an expert on the topic. It was only a matter of time, he said, before I found another one, and I had to be careful because it was just a phase—they weren't really my people. A Maniac Girlfriend, by his definition, was a beautiful vacation from the pressures of logic and world order: instead of planning the colors of your future wedding napkins, Maniac Girlfriend would tell you about how she could predict hurricanes, speak dolphin, or see through cement. Your relationship would last for any length of time between one hour and six days. You would never see Maniac Girlfriend again, but if you did, it would be because she needed money.

Maniac Girlfriends were another creation of our permanently adolescent city, Jack informed me, just like all-hours delivery pizza. Of course, over the years to come in New York I forgot the most important parts of his advice at least a dozen times.

By age twenty-seven I had gotten into a fair amount of trouble with Maniac Girlfriends. At one point I tried to suggest to my latest tryst that maybe she slept with me because she was a little bit bisexual, but that conversation did not go well: it turns out that the only expert on your boss's sexuality is your boss. I realized I needed to confront the fears founded at the church fair years ago and agree to look for love among my people. I started spending a diligent amount of nights at a West Village lesbian bar called The Cubbyhole, where the drinks arrived in plastic cups and the ceiling was inexplicably always coated in streamers. The Cubbyhole's overall aesthetic was dirty piñata—the place looked like it was designed by an inebriated gay man stumbling home from a night of karaoke, desperately searching for a place to throw up. In my book it was hardly the environment to inspire romance, and I always left The Cubbyhole alone. I became mildly depressed. I didn't know how to transition from being the me who lived in New York into being the me who had a life in New York. The city didn't seem to care that I was struggling because it was always moving on, always alight with the new next thing.

One night beer and I decided to sign up for Match.com, seduced by an effectively placed 2 a.m. television advertisement for the service. Match.com was a lot like putting Rosie from the *Jetsons* in charge of finding your significant other: it was mechanical and overbearing in a vaguely maternal sort of way. Match.com's profile creation process was littered with politely insistent, italicized advice: *"Think about what you want your pictures to communicate to others. Are you friendly? Are you happy? Hint: if you post a picture of yourself with a smile, more people will respond positively to you."* What a way to phrase it! I could easily imagine Rosie the Robot's original draft of her same thought: *"Come on, fatty,"* she'd have growled, covering her burp with a cigarette, *"we've got research on this stuff. Don't you go trying to reinvent the wheel. Put down your Stouffers French Bread Pizza for once and smile for the goddamned picture."*

I set out to communicate my own message. My profile photo, I decided, would be a snapshot of me right next to the gold Charging Bull statue on Wall Street ("Financially minded!"). Said picture was taken well after midnight. Drunk, I am kissing the bull's butt. In the background a man on the street is laughing at me and pointing me out to another man, who is laughing at me as well. Once I had decided on the profile picture, I added in some groovy lifestyle shots to round out my visual story. First was a picture of my best friend and me by the ocean in the middle of a severe hurricane; we are dressed as oracles and hovering over a Ouija board ("Prepared for anything!"). Then came a picture of me making a fire ("Marketable skills!") along with some kids atop an abandoned building in Chinatown. The fire picture captured the tragic moment when I learned that my friend Drunk Karen spaced out and forgot to bring hot dogs, but to the untrained eye it looks like I am simply planning my next move.

Next came a barrage of questions apparently designed by computer-wielding grandmothers, software engineers, and social studies teachers. "What are your life goals?" "Do you like movies?" "What religion do you practice?" These questions were about as exciting to me as meeting at Denny's and laughing over the inaccuracies of old maps. Plowing through each one and its lazy-person-eliminating, two-hundred-word-count requirement was a supreme exercise in patience. Mid-effort my friends texted me several times, suggesting I abandon my online dating idea and come out partying with the lot of them. I could look for someone to love the old-fashioned way! A longtime ally of the computer's, I adamantly refused these offers. How could continuing to fail at things the "old-fashioned way" be nearly as effective as the new ruthless efficiency of Rosie, my online dating dominatrix? How many other people was she yelling at to finish their profiles at this very second? Nothing could mathematically compete lest I walk around in a T-shirt that said, "I'm looking to date! Ages 28–40, must be a female nonsmoker and not give one shit about Jesus."

It was dawn before I was done with my dating profile and officially approved by the authorities of Match.com. Who knew what would come of my life now! I uncorked a bottle of wine and punched my computer in the face. Making an online dating profile was the worst thing ever. 1–800-Flowers should create a line of sympathy baskets in honor of the occasion.

The next day, having officially walked through Match.com's gates, I saw something I'd never experienced before: the equivalent of New York at 6 a.m. with the lights flipped on as everyone sobered up and faced reality. Match.com was a bunch of otherwise urbane and high-nosed adults sitting in their chairs and calmly describing who they were, how they lived, and what they actually really wanted from their future. The resident Maniacs of the city were nowhere to be found; I suspected they were why OkCupid was invented. There was nothing joyous about the exercise of looking up people to talk with and then thinking up things to talk with them about, but it felt both cleansing and necessary. My worst day on Match.com was far more earnest than my best night in a bar.

Months later, on one disgusting winter afternoon when it was storming in a way that seemed like the sky was having an accident, I was sifting through my Spam folder when I stumbled across a junk mailing from Match.com. It was Rosie, wondering how I was doing on her website. *Hello [user name] Jess981_360,* the message began with a personal touch, *We see you haven't been around in a couple weeks. See who you're missing out on! Sign on to Match.com, and you could be dating Lauren675253!* Below, the website promised, was a photo of the real live person from Brooklyn who truly was Lauren675253. This woman's beauty was ageless in a way where it looked like she should be wearing pearls. She was playing a bongo drum in her profile photo, always a positive sign, and wearing a long, flowing dress. There was a little boy at her right who was looking up at her like she was the sun. What was this woman doing here in the Filene's Basement of the Internet, posting her profile on Match.com? How had she escaped being anointed queen of some modestly sized European country? I didn't care to figure out the answers to my questions—I had to make Lauren675253 mine. "Stupid computer!" I shouted out loud. "I want to marry Lauren675253, not date her!"

A month later Lauren675253 and I were sitting over plates of Italian food on Bowery and Bond. She was every bit her pictures. I was just trying to avoid the more massive cloves of garlic, not fall off my chair, and play it cool.

At first we had the obligatory conversation about being embarrassed to have met via Match.com. "It's horrible, but I didn't see another way to meet someone new," Lauren confessed, alluding to her son. I asked what it was like to have a kid.

"Parenting's, well, I guess the best way to explain it is that it's just more life," she told me. "Last night I was making macaroni and cheese out of the box when I bumped into the cat and said out loud, 'Oh my god, I completely forgot that you exist!' And I meant it! The poor thing, his water bowl was empty. I think he was surviving solely on condensation."

I laughed loudly, my mouth in an oblong O. Lauren laughed back. When I reached for my napkin she leaned in to kiss me.

Three minutes later the gravelly gentlemen seated at the table next to us sent over a complimentary round of Johnny Walker Blue. "Here here!" New York seemed to be saying to my kissing Lauren, *This is your best idea yet.*

We were married two years later. The funny part about our happy union is all the ways we could not be more different. Lauren is a person whose perfect bliss is defined as participating in a spirit workshop with a gentle, middle-aged man who's written a book-on-tape about finding inner peace, and I am a person who once cried tears of joy at Barneys upon spotting a markdown on a pair of leather pants. We are two people who, for our various reasons, were eased into our New York adult existence by a semithoughtful piece of junk mail from a really crappy website.

Our story is as modern and New York and Internet as they come, a riot to reveal. Seven years ago my wife had a baby when she was married to her first wife, and then not long after, they got divorced. Then my wife fell in love with me, her ex-wife fell in love with a handsome Parisian, and now all of us live on the same street in Brooklyn.

The good news is the part about Brooklyn, where it's perfectly normal for a kid to have one stepdad, one stepmom, two moms, two step-siblings, four homes, three pets, and a personal art teacher. In New York you can attend a fundraiser for your kid's private school (which owns an actual Picasso but is still constantly running out of money) and say, "This is Beckett's mom, and this is Beckett's other mom, and I am Beckett's stepmom," and everyone will shift their Chardonnay to their left hand and say, "Nice to meet you!" before going right back to quietly guessing how much money you make. In Brooklyn being a weirdo is driving the speed limit. I blame it on our borough's origins:

Manhattan got Alexander Hamilton, reinventing democracy, and we got Walt Whitman, who was fired from his job at the printing store because he couldn't shut up about flowers. We Brooklynites are not well known for our ability to focus. This is one of the many reasons I belong here.

A look at my home, thanks to the Internet: the kitchen cabinets are filled with enough goods for a hurricane. The cat is asleep inside the piano next to a crusted circle coating of barf that will take me hours to remove. On the deck the dog is trying to dig a hole to the neighbor's yard and appears alarmingly close to succeeding. My wife is huddled in the closet on a conference call, trying to create an artifice of quiet for the associate on the other end as Bennett fake cleans his room while screaming the lyrics to a song by Katy Perry. Our whole crazy family is due to leave for a preschool art event. No one is even dressed.

I turn the key and find my wife on the floor. She puts her phone on mute to look up at me and says, "Thank god you're home!

Friends Don't Let Friends Start Startups

It used to be the great American novel. Then the great American screenplay. Now it's the great American business plan.

—eDreams, 2001

MANY times people assume that because I love computers, I must love startups too. Those people are very wrong.

I didn't used to be such a grump. In fact, years ago I actually considered myself to be an advocate of the startup. I downloaded new mobile applications and attempted to work their feature sets into my daily existence, no matter how tangential it was or how much that app was *not* simplifying my life. I felt "excitement" when I attended a dinner party and found myself seated next to a CEO of a nascent startup. I believed, in earnest, that maybe the solutions to the world's most dire problems were just three dozen or so "killer apps" away. The passage of time and the continued existence of BuzzFeed have changed my opinions of startups, and I can no longer tolerate the querulous drone of a Caucasian male Millennial speaking to me of his startup's grand vision. If I see a T-shirt with a logo, or hear the words "secret alpha," or smell a customized bumper sticker adhered to a late-edition MacBook Air, I will hide in the bathroom for as long as it takes to escape that confident young person as they pitch something that does not need to exist.

By the year 2011 startups were fully cooler than they'd ever been in the year 2000, and all of my friends were leaving their regular jobs at normal companies for fun, cool, job-like objects at startup companies. Instead of marching into work each morning, these friends gathered in living rooms and asked, "Did anyone want coffee or food or anything?" And then once that was taken care of, they bitched about how hard it was to start up a startup. Several hours later someone said they should settle down, and then they started to work on their "idea." Today those same friends have no idea why they made the choices they made a few short years back, which is why I'm not asking them to tell their stories to you. Asking a person who worked at a startup about "what it was like to work for a startup" is like asking your friend who got blackout drunk to detail their thoughts on the later portion of their Saturday evening. The far more reliable witness, in my opinion, is the *friend* of the person at the startup, the boringly employed observer witnessing the hideous spectacle from afar. If that friend is a remotely decent person, at a couple of crucial junctures they attempted to talk their friend out of the startup, advocating *any other* kind of job or activity in its place— even grad school, even a useless MFA. There are several stages of the entrepreneur's progression into the pit of hell that is a startup company, and I am here to be your guide through each one:

Stage One: *So You've Got a Friend Who Really, Really Likes Startups*

⏱ **DURATION:** one to four years

☄ **DANGER LEVEL:** When you hear noises in the walls of your apartment each night but you're still able to convince yourself that those noises are definitely not mice.

Working at a company is so lame and boring, your friend Rory or Chad is saying! There are *so many* things that make companies so lame to work at, like benefits and office supplies and a boss who gives you regular feedback. Wouldn't life be much better at a startup? Startups are in business magazines. People who work twenty-four hours a day at their startup feel much more freedom in their lives than people who work at nine-to-five jobs. Just like what's his name, Horatio Alger. That guy would definitely be the cofounder of a startup, along with an engineer named Boris whom he'd never met in person because Boris lives in Russia and also maintained a side business as a club promoter.

Dangerous influences for your friend at this stage include seeing pictures of the Google cafeteria, reading the poetry of a then-unknown twenty-something Jack Dorsey, and/or meeting other people who also have ideas for startups but have not suffered the isolating hell of actually forming a company themselves. All of these things will solidify your friend Rory or Chad's beliefs that it's a good idea to build a wooden spaceship bound for the moon, also known as a startup.

Stage Two: *You've Noticed That Your Friend Has Bought Some Books by Famous Venture Capitalists That Are Full of Advice About Startups*

⏱ **DURATION:** One week

🚀 **DANGER LEVEL:** SpaceX ship bound for Planet Delusion

Once your friend is really wading in deep, it will be time to start reading the books and tech blogs of startup land's most revered figures: the venture capitalist.

A venture capitalist is like a member of clergy in the Catholic Church: he uses his power to prey on young boys and then forces them to keep the particulars of their relationship a secret. (Examples of startup-founder-and-venture-capitalist special secrets: this startup will never make any money / this startup is technically designed specifically and only to lose money / this startup has lost even more money than it wants to tell its investors about because it is just *that* much money.) The venture capitalist also expects patently ridiculous things in return for their involvement in digitally savvy young men's lives, namely a return on investment of "100x."

100x, in venture capital industry parlance, means that a good startup ideally will return *one hundred times the amount of money* originally invested into it. Would you ever take a loan out if you were expected to return *one hundred times* the value of that loan? Of course not. "10x" is also sometimes considered to be acceptable expected returns, but it is nowhere near as great as 100x. (Don't believe me? Google "Marc Andreessen 100x." *Pages* of quotes will appear.) 100x and 10x are two of venture capitalists' biggest and, therefore, best ideas.

The other thing venture capitalists have glamorized—almost as much as the idea of 100x—is a new, more positive emotion around the notion of "failure." Success is success, sure—that's why we have the whole 100x thing—but failure is success as well. Failure, you see, is a great thing for all companies to do because even though it loses them money, it teaches valuable lessons. For a while venture capitalists were unique in thinking this way about business—most regular businessmen didn't *aspire* to fail—but by 2013 or so, the venture capitalists had sent so many books, blogs, Tumblrs, and tweets farting through the sky that they managed to convince the world's *real* business people that failure was a good idea for *their* companies too. Publicly traded, massive, actually logical, actually-making-things companies were now aspiring to fail. Around this era it became common to meet normal adult business people and have them mention that they were really trying to encourage their team/company/division to "embrace failure," "fail fast and often," and to generally "fail more." The interesting thing about the type of failure that these venture capitalists lionize is that it's not real failure at all. Not in the way my grandfather's generation would have classified a personal failure. Not the kind failure where you kick your rented tractor out of frustration and it lights on fire and burns down the neighbor's farm.

The gods of the venture capital world are a pair of soft, glistening eggs named Marc Andreessen and Ben Horowitz. The highlight of my literary life thus far was calling Marc Andreessen a "Brobdingnagian Sloth Fratelli overlord" in *TechCrunch*, a publication that his staffers may in fact read. Marc Andreessen would like to be remembered as the man who brought the world Facebook (and Marc Andreessen), but I would like the world to remember Marc Andreessen as the man who brought us Fab. Other classic Andreessen-Horowitz don't-miss business mistakes include but are in no means limited to Quirky, Clinkle, Rap Genius, and Twitter.

But don't judge a venture capitalist by his mistakes because everyone knows that 99 percent of startup investments are failures and then there's always one investment that delivers 100x, just like it's all been perfectly calculated to work. We are talking about precision investing instruments here, not old white men who'd hurl money at any child with a PowerPoint.

Mr. Andreessen's business partner, Ben Horowitz, lists himself as a connoisseur of both barbecue and rap music. In fact, Mr. Horowitz's African American friends have called him "the Jackie Robinson of Barbecue." If you're scratching your head and wondering about the source of that stupid anecdote, you can find it on page one of his book. (I assume I don't have to tell you that at the time of this book going to print, Mr. Horowitz and his billions of dollars have funded a sum total of *three* companies with African American founders. What a guy! Are those the same friends who complimented the barbecue?)

Ben Horowitz's book is *The Hard Thing About Hard Things*. As you may have guessed by now, the text is not about ending racial injustice, fighting political occupation, or improving our crumbling national infrastructure; instead, *The Hard Thing About Hard Things* is about how challenging it is for well-educated white men like himself to start new software companies while working alongside other well-educated white men, aided only by all those men's collective assets, connections, experience, and political savvy. The jokes just write themselves. The jokes, however, were apparently not so obvious to the *New York Times* Book Review, which decreed the text to be "lively and candid," though later noting with a hint of sadness that the text's lessons were perhaps a bit "well-worn." The *Times* reviewer should be forgiven, however, because it is sad when our nation's plainclothes billionaires let us down.

Back to your formerly normal friend, though, because that friend is absolutely devouring these books, highlighting the shit out of the pages of *The Hard Thing About Hard Things*.

Stage Three: *Oh God, They've Narrowed It Down to One Idea*

🕐 **DURATION:** six weeks

💣 **DANGER LEVEL:** Taco Bell "Fire" Hot Sauce plus the all liquid garbage from an outdoor music festival

In the earlier stages Startup Friend will entertain several equally useless ideas for new mobile app companies. Let me try to guess for you right now what they are: so one is about food delivery, right? A specific snack they like when they're high? And one is about a very specific problem that they experienced while traveling ("I was flying to Indonesia last year when I noticed that there was nowhere to buy GMO-free insect repellant"), and one is about sharing fashion with like-minded strangers. How close am I right now?

In the Narrowing It Down To One Idea stage, things get dicey. An argument for all choices—quitting work, gambling on a startup, the hobo's logic that their startup will end up in the less than 1 percent that are successful—will not be successful at this stage. You've already lost your chance for at least a year.

Stage Four: *Time to Write a Pitch Deck for All Your Lucky Future Investors!*

🕐 **DURATION:** endless

💣 **DANGER LEVEL:** Gravity has ceased to exist.

The traditional venture capitalist blends the confidence of Alexander the Great with the intellectual firepower of Barney the Dinosaur, and so for this venture capitalist to "analyze" a startup company for potential multi-million-dollar investment, they must review a certain style of written material. One that suits their limitations. Said VC-friendly written material is called a "pitch deck," a PowerPoint that's generally in the one-hundred-to seven-hundred-word range, length-wise, plus unlimited graph charts. It's sort of like being at Olive Garden, only the food is bad ideas. The pitch deck is something your startup friend will agonize over creating.

The purpose of a pitch deck is for the startup founder(s) to select a centuries-old industry that they plan to "disrupt" by means of their digital "innovations." *Food is a $90 trillion industry*, a typical pitch deck will begin. *And by our calculations, if Food.io can capture even 10 percent of that industry, it's a $20 billion opportunity.* Simplistic, highly designed pie charts will accompany these decrees, with all the data points very creatively sourced. ("According to the website Chartbeat, the attention economy of the food universe is growing by over 96 percent!") Later on, in the section labeled "Our Team," photos of young men jamming out at Coachella, standing next to surfboards, or skateboarding down a public staircase will accompany their biographies, with these photos intended to increase an investor's confidence in the group's overall money-making capabilities. Somehow this works.

By the time you've read a dozen of these presentations, they go from being comical to the new normal of business communication. Startup pitch decks' most definitive trait is that they reduce profoundly complicated macroeconomic situations into folksy binaries, making any system's problems seem surmountable by a small team of mobile software engineers. In a typical pitch deck a problem-solution statement will go something like this:

Finance is broken.

Followed by

But Moneybong can fix it.

You try to see the good intentions in these presentations, at least for the first couple of years. You try to believe all the hype. *Young people want to change the world with their companies*, you think to yourself. *Isn't that nice?* However, the flaws in the optimism become ragingly apparent somewhere around years two and three of hanging out with your friends who have committed themselves to starting startups: your sleepless, constantly worried friends who believe they've deftly avoided the hell of working for a big company all while failing to recognize that the pressure placed upon them as a startup person may indeed be killing them. Is finance broken? Yeah, sure. But is the team at Moneybong really the exact assortment of minds that the world wants guarding its retirement funds? And can the whole of "finance" really be "hacked" by a team of children in a coworking space who have never so much as applied for a car loan?

Stage Five: *A Tech Team Has been Assembled (Sometimes in New York, Usually in Kazakhstan) and a Product Has Been Released*

🕐 **DURATION:** one to four years

💣 **DANGER LEVEL:** Your kid's pregnant hamster is hiding somewhere inside your apartment building.

The product is done! The product is done! Hasn't the whole world been waiting with bated breath for this exact kind of product to come out? By the way your friend is talking, you would think so. The email flow at this stage will be daily. Your friends will want you to promote their product and believe that you "owe it" to them to promote said product. Because if you had a product, they would definitely redirect their lives to promote you.

This is when the blogs will happen.

Your friend will get a write-up in *Business Insider* (which, by the way, would do a write-up about a write-up about an Instagram of a girl in a bikini at a South Carolina fair) or the *Daily Dot*. When it comes out, your friend, by their own estimation, is now well established, and the excitement generated from this one twirl around the ball with the press will be enough to fuel their startup delusions for at least another year.

This is the time when your friend will get a big ego but believe they are "staying humble" as they now carry no fewer than three different cell phones everywhere they go. The myriad cell phones will exist to both test the product on different hardware and also to field numerous calls from powerful people. One of my startup friend's signature hang-out moves is to casually-not-casually lay down their three to four cell phones on the table and then sit down behind them, like a king in front of his bowing subjects. Once I watched an actual doctor witness this ritual and then shrug ever so slightly with understanding: *Hey, that guy has to do Internet*, his inner monologue seemed to say, *I'm only a doctor.*

By now your friend is giving aggrandizing fictional business interviews to outlets like *Fast Company*. He imagines writers for these publications will lob him questions about the future of the world and the economy, and he will have answers—confident ones. "You know, *Fast Company* reporter," your founder friend might say, "our company is not just in the business of making _____. Soon we see ourselves extending our brand into the worlds of nutrition, education, hotels, oncology, birdwatching, and cleaning the water supply." *Fast Company* will breathlessly cover your friend's delusions of grandeur, rendering it business innovation fact in the minds of the reading public simply by committing them to paper: *"_____ startup is not just about _____; they're the future of the entire world."* In startup land the competition for making the most insane claims is extremely stiff. Uber: any time a person, place, or thing moves anywhere in a vehicle, we own that moment! WeWork: we built a utopia!!! Airbnb: we want to rent that utopia to someone else! Slack: beep borp beep borpp #@(((!)!!!!!@(@$!)!!!!

Secretly, by this stage, no matter how much capital your friend has raised, they are running low on it by now. They are starting to get nervous, and all the stuff they blathered to *Fast Company* was just an attempt to win some venture capitalists for a few more days, give them something to tweet to their fans.

Stage Six: *Your Friend Has an "Office"*

⏱ **DURATION:** one year, give or take a year

☄ **DANGER LEVEL:** Low, because this will drain the money from the startup venture capital coffers and potentially return your friend to you

Office time is always the time when your friend will have peak employees. The group will generally behave as if they know they are cool and fashionable but don't know they've just been recruited to a cult. Everyone will be very photoshoot-ready and monotheistically hip at all times. Conversations will be structured entirely around in-jokes. The group will begin wearing high-end T-shirts bearing the company's logo as their new default weekend and workout uniform, and shortly thereafter matching water bottles, gym bags, and baseball hats will appear, at which point the memorabilia will migrate to being a full-week uniform.

When young people are allowed to design their own offices, awful things happen. I've never worked at a startup, but I have worked at an office designed for and by young people, and that was pretty close. Every employee had a locker with his or her own name written on it. This office smelled like toast, dead rodent, candy, perfume, and burning plastic. Each employee had approximately two feet of personal working space, and then someone else's working space began, which of course made it impossible to work. There was also an office dog named Lola, who was a small, *Wizard of Oz*–looking animal acquired on a drunken impulse, and she was decidedly not house trained. Lola smelled like wet bacon and ate a steady diet of cigarette butts and stickers off our office floor.

Sometimes Lola barfed and then ate it during conference calls. It was impossible to make points while Lola gagged loudly, producing a blob of what looked like Campbell's Chunky Soup and then, cautiously sniffing at her own creation, would begin to lick it. We were begging her not to do what we could see her considering next, muting the phone and loudly yelling. Children, inhabiting an office.

The other thing I remember most about working in a startup-like office was just how many days could go by without anyone doing any work. Points and debates that I can recall from this office include:

"I can't believe they let newscasters fly a helicopter *and* do the weather. Does anybody remember the 1980s, when Jane Dornacker flew into a bridge?"

"Oh my god, I just sprayed Raid on my glasses."

"I think I have psoriasis of the liver."

"Does Zicam even work? What the fuck is dissolving inside my mouth right now?"

"I'm glad we've given Guy Ritchie a twenty-fifth chance at directing."

"Picture this game: it's a blending of Twister and Battleship called Battlestrip."

Because it's impossible to work in young-people offices, often nothing gets done. And for startups that's when the money trouble starts. And ultimately money trouble is a problem that's impossible to stop.

Stage Seven: *Reality Sets In*

⏱ **DURATION:** Quick yet long at the same time

♦ **DANGER LEVEL:** Of course having a bunch of debt is stupid. Of course it is.

Funding is running low. Very low.

This is because "venture capitalism" is just a fancy name for "debt." The Silicon Valley/Alley response to running low on funding is simply to spend more of the existing funding on some "big ideas." (Read: things that have not been thought through, which makes anything seem big. Dave Morin and the team at Path thought they could steer their dying social network toward profitability by selling people digital stickers. That's what I mean by "big idea.")

At this point the final stages will happen in rapid succession, sometimes within the constraints of a single week. They are: Back out. Never apologize to anyone for all the downloading, tweeting, emailing, Facebook liking, reviewing, and using that you made everyone do.

And finally . . .

Stage Eight: *Good-bye*

Every time a startup dies, a talented young person has to reprioritize. And to these young people, I say, don't become a bald, old white man standing over a barbecue, thinking you're a part of the solution.

2015
TECHNOLOGY
AND BEYOND

We live in a technological era so vast and so wonderful that you can ship your worst enemies glitter. (Don't believe me? Pick an enemy and go thank the kids over at ShipYourEnemiesGlitter.com.)

People complain about the Internet, but I have noticed that those are usually the people who forget that not long ago New Yorkers used to throw their poop out in the streets every night. The city burnt down to the ground on the regular, and worst of all, the shower hadn't even been invented yet. You couldn't even destress from your day of all the poop and fire with a nice long shower because it was, like, four seconds ago that someone thought up the concept of the shower. Even the rich people—those whose offspring would have populated a "Rich Kids of Instagram" of their day—their families passed down terrible career choices, like being a goddamned whale assassin. All those rich people who lived in Martha's Vineyard just a few generations ago had to murder whales every year to afford the place. Not the most Instagram-ready material. Now is not perfect, but comparatively, we're okay.

The moments in which I've loved the Internet the most have been the humblest. They've been the moments where I am sitting in some well-intended hospital waiting room, looking for a laugh in between pacing, crying, and bargaining with God. In one such moment I can remember stumbling upon a Wint tweet. Wint, if you don't know him, is a true genius of the Internet; he is an anonymous writer who's seemingly co-opted part of the human meme that is Jack Nicholson and mixed it with postmodernism and acid:

wint @dril
my watch beeps whwich means its time to stand in front of my ex-wife's house and play "Hit THe Road Jack" while dacning and licking her mail

wint @dril

i am selling six beautfiul, extremely ill, white horses. they no longer recognize me as their father, and are the Burden of my life

wint @dril

"jail isnt real," i assure myself as i close my eyes and ram the hallmark gift shop with my shitty bronco

Any one of those remarks could send me reeling on a normal day. But on a terrible day they are nothing short of miracles. They are what allows a person to go on for another hour. And they are available for free. On the Internet.

Adopt a Microwave
on Petfinder.com

I WAS innocently browsing the pages of Petfinder.com when a photo induced a full-on personal catharsis: I had found her. I had found the rescue dog my wife and I had never once discussed us getting.

According to the Internet, my future dog was from a litter of ten puppies that were rescued from a kill shelter in rural Alabama. Built like Photoshopped Labrador retrievers, these half-deer, half-dog forest creatures held my heart in their grip. After an hour of reading and rereading the same one hundred words of information, I selected my favorite puppy from the onslaught of photos and emotionally made my commitment. This was our dog.

Somehow, through this Internet photo, I was able to assess everything I needed

The very first picture of our dog, Waffles Karl Leslie, also known as Peanut, A Very Special Lady, Booful Girl Pretty Smile, and Stinky.

to know about this little dog: that she was a very good dog, very sweet with children, smart, a comedian, and for certain meant to be ours. I would wheel her around in a cart in ten years if her health required it.

"Honey," I called my wife at work, hoping the landline would make the call feel officious, "there's someone on the Internet I really want you to meet . . . " I was closing my eyes and smile-praying through gritted teeth. My wife is the person in our family who remembers to file the taxes; I am the person who once flooded the bathroom with lavender-scented bubbles because our son said he "needed a moat." My well-intended decisions frequently create drudge work for her, and even though I felt bad about this behavioral pattern, I recognized that this was not the day for change. "I found a junkyard dog from Alabama on Petfinder.com, and I love her."

I forwarded the photo of our dog and began talking. "I'm going to do ALL of the work of owning a dog—don't you worry," I said preemptively, wishing I'd unloaded the dishes a few more times in the last month . . . or ever. "I don't know if you remember this, but I had a dog as a kid—well, three dogs actually—and I was totally the person in my family who did all the work." I had now assumed the voice and rationale of a preteen. It was an obvious exaggeration, but I was desperate. My best friend was locked in a cage in Alabama, and I needed to get her out.

Lauren sighed. From my end of our phone call our twelve-year-old Himalayan cat began repeatedly throwing up in the background. I watched as bright green bubbly foam emerging from all sides of his mouth—a Wylie Dufresne–esque tribute to the leaves of our houseplants—and as the bright, bile-y goo sunk permanently into the fibers of our living room carpet. That was the third IKEA carpet we'd purchased in the last four months alone. Before our beloved cat Leo turned one thousand in people years, we'd owned a very glamorous white carpet from a store called Design Within Reach. The extremely expensive store's name is a true New Yorker's in-joke, but I'd decided to patronize it anyway, to prove a point to the world and no one. Once the cat swiftly destroyed the beautiful rug on which I'd spent over $700, we downgraded to a series of West Elm rugs, and after that to a CB2 rug, and then lastly to a succession of IKEA rugs, as there are no more downgrades left. Within one year of living with chronic elderly cat vomit, we began reordering the same humble FARFLATTT or whatever $65 plastic "rug" that IKEA seemed to manufacture specifically for the parents of things that floor barf. *So you can't have a nice carpet!* their smug, homogeneously ageless Swedish faces seemed to be nodding in our direction. You could feel the clipboards in their arms, having thoroughly studied our situation and arrived at a textile conclusion. *Well, here you go, tacky American. (And we know you're an American.) We made you this shit, the—what did we name this thing?—oh yeah, the FARFLATTT. In Swedish that word translates to "hostile inside joke."*

Well, we already had the carpeting for a puppy, I thought to myself, justifying the considerable pressure I was putting on my wife, who, at the time, was not a dog person. Growing up, Lauren's family had a stray cat named Taffy, a mean cat who loved no one else in her family except her. That was how Lauren preferred her animal relationships: exclusive and committed. "Lauren, I love you so much, and our doggie will too!!" I shouted spastically at my wife, hoping to cover up the sounds of the preparatory gagging from Leo. I shot the cat my most vicious face of disapproval. I needed the animals already residing out in our house to seem like they had their acts together, and although the hermit crab was maintaining the peace, our asshole cat was doing a great job demonstrating just how problematic pet ownership could become.

Despite the barf, my techniques seemed to be working, as I could sense that Lauren was slowly opening herself to the concept of rescuing the dog I found online. It *was* a pretty cute photo, she had agreed. We proceeded to have a thirty-minute discussion about day-to-day responsi-bilities, medical bills, dog walkers, contingency plans, and house train-ing. I made it through without revealing my ignorance on all matters. I was allowed to email the dog rescue and schedule our home visit.

At nine o'clock the following Saturday morning my wife answered the doorbell, and a sunny little baby deer noiselessly zipped through our apartment on legs that looked like stilts. After completing her inspection of our domicile and crashing down on the rug, exhausted, the puppy looked up at us with her crooked ears and omniscient, amber eyes. She was so helplessly cute that in the course of two minutes she'd won over even Lauren, who had gone from stressed-out future-parent pacing to pressing her face deeply into the little deer's golden belly.

Deer legs, question-mark tail, and semi-tiny head.

"She's a baby!" Lauren cried, surprised to be so enchanted with this dog-creature in front of her.

"She's a real pleaser," her foster mom told us. Having grown up on a farm, Foster Mom's animal pronouncements carried considerable weight.

"She looks really smart," I added, attempting to seem on the same plane as Grown Up on a Farm Foster Mom.

"I will warn you," Farmer Foster Mom said in a tone intended to separate herself from me. "This dog? She's a little needy. She likes a lot of attention." This was an incredibly cute animal, Foster Mom implied, but *not* the kind of animal who would make it back on the farm. Luckily our new pup was sprawled out on the couch in a two-bedroom apartment in Brooklyn, and no practical survival skills would ever be demanded of her again.

"She's a Leslie!" Lauren decreed, now fully a different person. Our dog, sensing the permanence of true familial acceptance, responded to Lauren's love by releasing a steady stream of toxic farts into the air. It was the canine equivalent of unzipping one's party dress and kicking off one's heels, grateful to not have to pretend any longer. Our canine family member was home at long last: Waffles Karl Leslie had finally found her tribe.

I should explain how Waffles Karl acquired her unique name. "Waffles" came about several years earlier as I riffed on Instagram-ready pet names one night with my sister-in-law, Erin. I had actually forgotten about the moniker, but Erin is the kind of person who will remember what you and margaritas wanted to name your nonexistent dog three years earlier at a midpriced Mexican restaurant in Carroll Gardens. "Karl" was coined in the wake of my mother's immediate disapproval of the name Waffles. "Dogs are supposed to have hard consonant names, or at least a sibilant consonant blend! A dog can't hear the difference between a name like *Waffles* and general ambient noise!!" My mom was confused as to how her scientist parenting had gone so wrong. My neighbor Angeline heard out both sides of this argument a few hours later; as a dog lover by hobby and a litigator by trade, her ruling would stand. "Karl," she responded, "that's a nice middle name. Very consonant." And Waffles Karl was born.

"She's a little puppy angel!" I bragged to my mom over FaceTime after we had been dog parents for a full two hours. I was dying to show off my puppy, a puppy that was, by my measure, already vastly superior to my family dogs from the 1980s to the 2000s, as raised by my parents. My mother had spent the decades rearing golden retrievers with a series of truly bizarre afflictions: one compulsively ate hair scrunchies, which made for festive poop; another spent her days in the backyard arranging rocks in circles, like a taciturn Paganist. The last one was so stupid that he would continually sit by the fire and cry because he was hot. My dad nicknamed that one "Our Last Dog." Last Dog ate pasta boxes and banana peels. Last Dog didn't know how to back up and therefore got stuck a lot—he'd cry desperately and wait for someone to move the particles of the universe in some mysterious way that would allow him to be freed from the tyranny of "between the couch and the coffee table." To understand Last Dog was to know that Last Dog would rather wait hours for help than adapt by way of finally learning how to back up. Instead of walking or running down staircases, Last Dog preferred to gracelessly fall down them; it was like watching a dog be kicked down the stairs by a ghost. One nice thing I can say about Last Dog is that because he habitually ate bars of Ivory soap out of the shower, he always had very pleasant-smelling breath. And he was a very gentle dog. But he was a supreme idiot. Perhaps his name had been some sort of cruel destiny: Last Dog's given name had been Phineas, after Phineas Gage, who was famous in the scientific community for being the world's first accidental lobotomy. Phineas Gage the person was a railroad worker who'd survived an iron rod careening through the frontal lobe of his brain; a similar incident might have well explained Phineas Gage Kimball's intellectual capabilities.

Determined to show my mom and dad exactly how this whole dog-raising thing was intended to be done, I was certain I could out-parent them when it came to a domesticated animal.

"Well," my mom said in that folksy riff-on-authority voice she sometimes uses to make points, "puppies don't show you the full range of their personalities until they're six months old. We'll talk then." There is nothing like a scientist to dampen the joy of an Internet purchase.

Weeks later Waffles Karl Leslie, in seeming collaboration with my mother, had managed to develop a series of bizarre character traits, including crying at an orca whale pitch for hours on end. After a late-night crisis in a rainstorm that basically involved me begging my dog to be good while she bravely refused for five hours to go outside and pee, I needed some people with whom I could vent. I decided to join a Brooklyn-area Facebook group for the owners of rescue dogs.

"Does anyone else's dog make microwave sounds?" I asked the Facebook group casually one Saturday morning, shyly but urgently making my debut on the forum. One person immediately wrote back asking for details on what exactly qualified as "microwave sounds." This was not a group that one could ever criticize for lack of attention. "Like, well . . . a microwave," I responded. "Like a dishwasher that's done with its work. Like a piano, if a piano had an extra couple of octaves tacked onto the high end and became capable of mimicking my dog's exact pitch. It's excruciatingly high, like what I would imagine sonar sounds like or other things human beings aren't supposed to hear. She may or may not be communicating with whales," I added, hoping that explained the severity of the situation. I imagined the butterfly effect of my dog, her calls helping a family of orcas in the north Pacific navigate their way out of the path of an approaching windstorm. Maybe she was not actually a high-maintenance, sweet, and needy dog but in fact an incredibly important player in the animal universe. The Facebook group, though incredibly fast to respond, was uncertain as to what I should do next other than to pay a trainer to come to the house and fix her. I wasn't going to do that, though! I was going to fix her with the healing powers of online advice! I "liked" each person's response to my question and dutifully replied to many of them, but I'd emotionally moved on.

We did the best we could with our dog, which is to say that we're the same as most other dog parents and their dogs. Waffles is extremely sweet, and she'll often sit down if you ask her nicely to sit down five to ten times in a row with a voice that grows slightly more ominous each time. She gets up on her hind legs and Russian dances for attention whenever we sit down to a dinner she likes. She's a protector dog. She once discouraged an intruder from his drunken attempt to open our back door while I was on a late-night conference call. (I didn't even have to drop the call—helpful!)

A few months in, once monitoring the puppy behavior was no longer a part-time job for every member of my family, I discovered that simply owning an animal in New York City and taking it to public places creates its own world of issues. Back when the island of Manhattan was traded on good faith to the Dutch for a handful of beads, it was best equipped to house a few peaceful tribes of Native Americans, not millions of people plus their three hundred thousand dogs. The city's layout all but encourages conflicts between dog owners. At the lovely, wooded Hillside Park in our Brooklyn neighborhood one morning I would encounter a stubborn father who'd proudly brought his barely mobile one-year-old human baby out for a walk along with his terrier. This one-year-old baby was not only allowed outside his stroller, but he was allowed to teeter through the packs of dogs with a snack in hand. A snack! And a croissant, no less! Most dog owners know that dogs, emboldened by a fenced-in park's off-leash status and the presence of many other dogs' butts, will proudly knock a kid or a fully grown adult flat for a snack. To a dog, a one-year-old with a baked good is just a stupid and odd-looking dog with a delicious toy. It was only a matter of time before this situation would end very, very badly. I waited it out for a couple of seconds, certain that one of the German shepherds' owners might think it best to intervene, but everyone sort of pretended that the guy and his butter-coated baby weren't actually there. It would be up to me to do the right thing.

"Excuse me," I said, wandering over to the kid's dad and his ironed Polo shirt, proud that I was the dog owner in the park who was taking the high road. "Not to interrupt your morning, but I have to offer this advice as a fellow parent," and I really let those last three words ring out because I felt they should justify so much. *As a fellow parent.* "Your son and his snack? They are not safe here. I've seen children get badly hurt in this park, I've seen EMTs come rushing in, pushing the dogs aside, parting them like a sea of useless goats in order to get faster access a bleeding child. Each time it happened it all started just like today. "

Okay, so almost none of what I said was true, but if you'd have seen the look on this dad's face, you would have realized that he required raising the stakes. Before I'd even said a word of my fake story, this dad was immaculately glib. He's already dismissed whatever was about to come blubbering out of my mouth. *He was judging me*, I realized. *Just because* I was wearing exercise pants, battered old Converse sneakers, and an ill-fitting GOP Teens! T-shirt that said, "@GOPTeens: Where should America's next war be? #GOPTeens!" on its front. The look was topped off with a baseball hat from a fancy resort I've never been to. This dad was judging me! Just because I *looked* like a crazy person! And I was also intervening in a situation where my opinion was unsolicited, the way crazy people often do! The nerve of this guy was unbelievable.

"The shirt's an Internet joke, by the way," I said, now realizing that the shirt might need a caveat. "I'm not actually a Republican, of course, I'm an extremely liberal Democrat who's taking the time to troll the Republicans, by way of wearing this shirt." The clarification did not benefit my cause. I realized that I had been wrong to judge Brooklyn as being free of pet-owning Republicans. I was embarrassed. But I was also still right enough to remain entrenched.

"Walter is fine," Walter's dad said to me. *Of course the kid's name is Walter*, I thought, growling to myself and wishing I had some of the dignity back that my outfit had robbed from me. Walter's dad looked like the kind of guy who'd expected his wife to split the check on their first date: like splitting a check was his idea of the beginning, middle, and end of women's rights. Walter's dad looked like the kind of dad who got upset if any of his things were touched, and his things were placed all over his very fancy house. Like a record collection of jazz music that he didn't actually understand. And a baseball in glass, something that should be a kid's toy, encased in glass—that was exactly the kind of guy that Walter's dad was. And he didn't buy wine at the liquor store like the rest of us—oh no, for sure he "maintained a wine collection." I was just not having this dude. And now we also had an audience.

"I come here every weekend, and nothing has happened," Walter's dad said to me, only a teeny-tiny bit swayed by the presence of the silent crowd.

"Well, I will certainly make a note of that," I said back, clipping my dog back on her leash and walking backward toward the park's gates.

"You do that," he said. Meanwhile his kid, of course, was also somehow miraculously surviving. Somehow twenty-five different Brooklyn dogs from different rescue organizations with different levels of mental disabilities were ALL simultaneously behaving themselves and not tackling this kid to the shit-riddled ground. As if any of these dogs were good! These dogs were kindly monsters! These dogs were normally disgusting. They were the type of dog that would leap up on one of the park's picnic tables and lick your coffee the second you weren't looking. They growled over the water bowl and peed on your shoes and knocked you down so hard you were left Googling knee surgeries. I was mad at all the dogs—except mine, of course. Mine was a good dog. All the other dogs were conspiring against me. A kid was teetering around with a high-end baked good, and they were just going to pretend he wasn't there?

"Have a nice afternoon," I said with malice, a little bit hoping it was curse.

"Yup," he said back. *That guy.*

And with that, Waffles and I left the park.

I later posted at length about the whole experience on my Facebook group, excited for the rush of the many people who would agree that I'd made the noble choice. I drew the story out for everyone's amusement, describing the knowing looks that I really felt I'd gotten from the other dog owners as I'd walked over to confront the guy.

"You were totally right!" my many new Internet friends agreed, many relaying stories of similar events in their own dog parks. "Good for you for standing up for poor Walter." "That kid's gonna learn the hard way!" chimed in a more outspoken member of our group. "Does he bring bacon to the zoo too?" A lawyer or two chimed in to outline what legal repercussions could hypothetically be, given the posted dog park regulations. I spent the morning going back and forth with my Internet group, and I felt like I was twelve years old again, whooshed back to the joy of my chatroom days.

We live our physical lives, if we are lucky, surrounded by the comfort of a few family members and friends who "get" the context of us. We drop and pick up conversations over decades; we experience the polar covalent bond that is someone understanding a reference to a joke that's almost as old as we are. However, where our friends' and families' knowledge of us can offer extreme breadth, there is a depth to be found in hyper-specific communities on the Internet that's almost impossible to replicate elsewhere. When you delve into a Facebook group designed for people from Brooklyn who have recently adopted a rescue dog, there is an immense amount of understanding to be found in what makes that life experience agonizing (saying sorry to your partner for the microwave sounds that are altering the life quality inside your tiny apartment), or wonderful, or comforting, like when they sleep under a blanket with your kid on a winter night. With just a single photo or line or joke, a hundred other people are nodding along with the full context of your experience. Such understanding often begets additional tolerance for the mundane. Case in point: my Facebook group will never, ever tire of photos of dogs smiling. We always have time for more.

When I was a kid my Internet friends were my best friends—my only friends, even. Because we shared one weird interest in common, I felt it was grounds for us to share everything. I was the fifteen-year-old who was so desperate to be a part of a bunch of forty-year-olds' lives. *I get what it's like to have a job!* I

That hideous face she's making here? That's what she does if you tell her to "do her pretty smile." It turns out that a lot of other people's dogs smile too. It's been a running thread in the Facebook group for at least a couple of years now.

thought to myself. *I work after school every day at the coffee shop!* The disconnect was an obstacle to overcome. As an adult I now treasure the fact that none of these rescue dog people will ever want to really get to know me. We'll stay "the one who always says the outlandish thing" and "the one who always says the sensible thing" and "the one who likes every photo first." The longer I'm an adult, the more I see the profound value in such limitations.

It turned out that I had effectively scared Walter and his father way from the Hillside dog park for the rest of all time, but new antagonists took their place in the form of two purebred Doberman pinschers. They were terrifying to me. These dogs' faces always looked like they had just gotten a secret message from the future that the apocalypse was nigh and the best thing to do was to round up all the humans into a circle and kill them so that the other Doberman pinschers could survive. Their tails were docked for efficiency, their coats sleek and aerodynamic. Comparatively all the other rescue dogs looked useless, like God's spare parts, the leftovers from his more focused efforts in making Doberman pinschers.

The Doberman pinschers and their humans spent every Saturday and Sunday from 6 a.m. to 10 a.m. "training" at Hillside Park. I nicknamed these dogs the President Obama Dogs because this couple was taking their training process so seriously that one would assume the dogs were being prepared for a task no less than protecting the president.

The best part about the President Obama Dogs was that after a few minutes of concentrated observation, a person would notice that both dogs were hilariously incapable of following even the simplest of commands. "Poseidon! Ares! Find the stick!" their whistles-on-a-necklace, married owners would shout almost in unison. Feeling noncommittal, one of the dogs would arch its back and poop. The other dog would trot off in the wrong direction and then investigate the underside of a very interesting picnic table. Hurl a tennis ball across the park, and the dogs would lie down. The only times when the President Obama dogs were capable of rallying any sort of coordinated response was when the universe inspired them to be total assholes. One day, for example, they decided to knock my kid flat on his back. Both of the President Obama Dogs suddenly ran at full, explosive speed straight at Beckett and then knocked him with such force that he flipped backward in the air and landed, like a spider, on his hands and his feet. Thank god the kid has always been hyper-coordinated. After I assessed that Beckett was okay and gave him many, many hugs, I knew that the President Obama Dog owners and I were going to have some sort of moment. I felt ready. More ready than I had been with Walter's dad. This time my T-shirt didn't say anything inflammatory, and I looked slightly less like the kind of New Yorker who brings a metal cart full of cats to the grocery store and is currently writing a one-woman show called "What

Doesn't Kill You Makes You Vodka" or something like that. I looked like a respectable human being, and I had a respectable reason to be upset.

"I am so sorry," the mom came over and said to Beckett first, who's never been a crier but looked especially tough after so adeptly surviving his surprise back flip. "Are you okay? Is there anything I can do for you? I am so upset at my dogs right now." She genuinely meant the words she said, and sure enough, her husband was already hauling their dogs away. Beckett and I felt bad for *her*. It seemed like Poseidon and Ares were in for a whole world of training as soon as they got home.

"I'm okay," Beckett said to Obama Dog Mom, and he smiled at her because he is truly a good person. Beckett is the kind of person who, at a mere eight years old, could see into her eyes and appraise her embarrassment, and his words were engineered to assuage her pain. The park's toxic, pee-glazed wood chips were still clinging to his shorts and hair and hands. He'd only just been kite surfing through the air a minute ago but had found the wherewithal to be nice. I couldn't believe it. And I knew there was only one group of people in the world who would understand the magnitude of what had just happened between me, my son, those idiot dogs, and their surprisingly kind owners.

"You will never believe what happened this morning!" I reported back to the Facebook group—and they got it, the whole thing. The highs, the middles, the lows, the feeling of having judged a person, been right and wrong all at the same time. They even knew the feeling of a dog as massively powerful as a Doberman pinscher knocking you over, that the breed was no joke. Within seconds comments started flying in, all of them praising Beckett's level-headed reaction, many of them caveating it with "I don't know if I ever could have been that nice myself!" Later in the chat others offered practical tips, like when you see a dog running toward you at the dog park, immediately bend your knees. I made an effort to respond to every comment.

What those of us who were first on the Internet in the late 1980s and early 1990s remember the most is the richness that this intense, specific camaraderie instantly added to our lives. There was a chat room where it wasn't weird to have memorized H. P. Lovecraft; it was basically a requirement. In such a chat room you weren't a loser who deserved to be tortured until you departed for the Ivy League; you were a leader in possession of a valuable skill. Your role would be judicial within that online group. As the Internet became a repository for all the things in life we were interested in but didn't always share in common with our physical peer groups, it started to find its place in more and more people's lives. Laugh all you want at America Online, but back in the 1990s parents who had long been Internet averse first got online to join a message board and seek out tips for their upcoming trip to Disney World. ("Bring your own box of wine—they charge a fortune there!" a helpful housewife chimed in on an AOL board my mom found back in 1998.) In the simplest of senses, enterprise services notwithstanding, that is how the Internet grew. How slowly people who had shunned the notions of email, the digital written word, and "typing someone a letter" were now revising their opinions to say "Well, I did use it for this one thing, and actually I learned a lot." And all the nerds smirked.

The time when Waffles ate Christmas was definitely a defining moment for our family. Waffles was almost a year old when we bought a Christmas tree, and we were by then experienced dog parents who could navigate the ins and outs of microwave communication and dog-park-parent social interaction. "Why would she do anything to the eight-foot-high Christmas tree that we just bought for almost a hundred dollars!!" we said to ourselves, knowing everything would be fine and we should spend more money to buy lots of lights. We spent the day decorating the tree while Waffles watched.

Waffles, stuffed into her Christmas sweater and plotting her big move.

My wife's family takes a kind of Raggedy Ann approach to Christmas ornaments: there are a lot of homemade things, school projects, macaroni. I add this not to criticize, as it's a lovely affect, but to foreshadow for the worst—can't mention a Raggedy Ann Christmas ornament in Act I without it being loudly barfed up in Act III. For the day of Christmas music and tree decorating, Waffles slept happily on the couch. She was even fine that night. Disaster struck when we left her alone the following Monday morning with the fully decorated Christmas tree—electric lights and all—and she decided to eat it. Waffles was waiting for one of the Cheeky Dog! Doggy Daycare staffers to come pick her up for the morning, and apparently sixty minutes of free time was simply too much.

I can't remember whether it was Christopher, Carver, or Kathy who flipped the wrong side of the coin and had to pick up Waffles that day, but the scene was gruesome. The Christmas tree—all eight feet of it—was somehow lying flat on the floor, the glass from the lights shattered across the living room. Investigations that took place that afternoon would reveal that many Christmas ornaments were AWOL. Not the new ornaments, mind you, not the easily replaceable ones purchased from local stores, but the 1970s ornaments. The ornaments made out of love and macaroni from Lauren's childhood. I was going to have to call my wife at work and let her know that the dog we bought on Petfinder.com had eaten all of her most precious memories. In my head I rewrote the Christmas present budget underneath the Lauren ledger. I picked up my phone, swallowed my pride, and opted to dial her landline.

Lauren, rightfully, was upset. When she got home a few hours later and opened the door, Waffles was sitting in her self-designated Shame Area, also known as the sixth stair on our staircase. Lauren started to laugh. Waffles looked away, toward the wall. Pieces of ornament crinkled beneath her feet. Eventually Lauren looked at me and genuinely said what I hadn't even allowed to cross my mind: "You've got to share this one with the Facebook group."

I didn't take a picture of the crime scene that day, which of course was a decision we now regret, but I did post a quick description of the general events. It wasn't live for ten minutes before the outpouring of photos of destruction scenes past started flooding in: dogs destroying various religious holidays, eating Passover feasts. Birthday presents opened with their stuffing gutted and displayed strewn across the living room floor as a child cried in the background. A robber dog who broke into his owner's barn and ate all the chicken feed. A dog who saved his family from a dangerous Roomba. Once again the pet Internet was leaping into action and brightening our day as I pinged back and forth with several of the dog owners. If you're going to rescue an animal from the depths of Petfinder.com, once you realize that you have no idea what you're doing and all of your valuables are now at risk, I suggest finding a Facebook group to share all the fun with. They're the perfect people to help you through that one.

Epilogue:

Seventh Grade
All Over Again

MOMS Google each other in New York. Sometimes the mom-to-mom Googling is for ridiculous purposes, like *Can that mom get my accessory line covered in Women's Wear Daily* and *What did they pay for that apartment? I can't believe I have ever touched someone's hand who has that much money.* Sometimes it's more sane, but either way it's a ritual. It took me a while to become one of these people, but eventually I did.

The thing about Mom-Googling is that it's a requirement for parent survival in the city. Say your NYC kid is forming a friendship with another NYC kid. Buckle up! Because that relationship is not only their weekend commitment; it's yours as well. New Yorkers don't have easy ways to get to each other, so a weekend play date means getting dressed; hauling ass to the subway because obviously you're late; figuring out the subway is broken; taking another, far less convenient subway; finding somebody's place in what is surely a different borough than your own; and locating their apartment with Google Maps and hopefully not making too much of a confused scene of yourself. None of that is relaxing. Because of the hassle, often the hosting parents will offer up some form of entertainment as a consolation prize for the adults, like cocktails or brunch. In the suburbs it's much easier to socialize because you can ride your own "personal subway" from place to place because you have what is also known as a car. If I had a car, I wouldn't need to Mom-Google; I would drop my kid at their playdate and blast off toward a blissful three hours alone at Barnes and Noble. But because I don't, I rely on Google to answer a few of my basic questions about other parents, which include:

Are these people that weirdo kind of New York industrialist who seems normal but after wine will blurt out that he/she doesn't understand "Why the homeless just don't go out and get jobs?"

Does the mom consume regular people food, or am I going to have to learn about raw food, Whole 30, or Paleo diets?

If Mom's a *special eater,* will she put out *her food* on the table, like sprouted nuts or whatever, or will she put out *food that she would never touch* and watch me eat cheese on a cracker like I'm a freak show imported from Coney Island?

Does the dad work in finance and viciously resent every second of the weekend that he's not able to spend in the office with his beloved quartet of computer monitors, reading Barron's?

If I try to make light macroeconomic conversation with Finance Dad, is he going to sneer at me like my pasty Irish face is eating away at his life force?

Things were different when I was a kid. In the Kimball household summer days meant only one thing: an afternoon trip to the man-made public pond that was just a couple of miles from our home.

Once properly cajoled, my mother would pack a mesh bag of all our beach essentials—Neosporin, butt wipes, Q-Tips, a role of paper towels, raisins, one good-flavored Quaker Chewy granola bar and two terribly flavored ones (which would guarantee a fight between us kids and give her an excuse to haul us home), buckets for catching water snakes and baby catfish and a magnifying glass for looking at them, peanut butter, stale graham crackers, maybe some pretzel Goldfish if we were lucky—and then we would pile into our family Volvo, which was less a vehicle and more a four-wheeled container for dog breath. My younger siblings and I would play in the lake's peanut-butter-brown water for hours on end, breaking only to spend all of our chore money within seconds of the ice cream truck arriving. Twenty minutes later my mother, chatting away with other mothers underneath the pine trees, wouldn't care that I'd recently barfed pieces of Blow Pops into my fishing bucket. She was hanging out with the Mom Crowd, and *they* were happily laughing at all of us.

Eventually the fun would end when my three-year-old little brother would succumb to the hours of furtive big-sister teasing and point, with a shaking finger, at me or Annie and let loose with one of most original child-of-a-scientist insults, such as "You . . . you one-celled organism!" That would be my mom's cue to roll her eyes and pack us up. She'd shout good-bye to the mom cohort over her shoulder in that impromptu way that young kids so encourage: *Time to go home!*

By the time I was twenty-nine and a stepparent myself, long gone were the days of kids swimming in liquid peanut-butter water while mothers idly bantered together, never thinking to take a photo, never itching to capture the moment for all of public posterity. By 2010 things were very different.

When you're a stepparent, you observe parenting rituals with a little extra lucidity. Virtually overnight you "step" into the role of being a parent, and you anticipate finding a bunch of adults on the side of that rainbow—you know, sanguine folks who know how to change the car's oil. Instead, you learn that thanks to social media, many of today's adults spend a lot of time engaged in the sort of preening rituals normally associated with teenagers. Modern New York City mothers are curating their online lifestyles, tracking their group's social elites, accruing followers, and developing an aesthetic. The computer has undoubtedly left its mark on the parent. We stepmoms—always fearful that we're going to become one of those heinous, Disney-created personalities worthy of a scepter or a cape—we don't know how to fit ourselves into these rituals. Are we also supposed to immediately participate in the parenting-on-the-social-Internet-thing too? Start taking black-and-white pictures of our stepkids wearing fedoras at brunch? Or floating faux-tranquilly in a pool after we've shouted again to "please close your eyes and look relaxed" while we get that solar flare *just so*? Or is it more respectful for everyone if we just opt out?

At first I thought it would be best to opt out, stay chill, and not make a big deal of things. I left my social media relatively untouched by parent world, instead a collection of my 95 percent childless work and life friends. These friends were still proudly chronicling their every Thursday through Saturday as if every day was St. Patrick's Day. You would wake up and just see a bottle and a costume for no reason. Given my blindness to the never-ending churn of mom-specific social media, I was more than surprised when, the morning Beckett started school in Manhattan, the *Times* had dispatched a reporter and a camera crew to interview moms about their specific takes on back-to-school style. Not the students—the *moms*. Their style had probably been spotted on some Tumblr or other aggregator, and then they were hunted down in their natural habitat: their kids' school. I watched in amazement as women wearing ostrich-feather coats and *Almost Famous*–style sunglasses wrapped their bangle-bracelet-clad arms around each other, expertly posing for photos and rattling off the names of the different designers they wore. While other clusters of moms stood in line for the photogs, they Instagrammed themselves posing together, looking like extremely thin and rich teenagers from Rachel Zoe high school. Their luxury goods crashed into each other; their pursed mouths showed off lipsticks in trending Pantone hues. I was dead from jealousy-embarrassment. These were not like the moms from Eastbury Pond. I was like the moms from Eastbury Pond. And like the moms from Eastbury Pond, I had not anticipated that back-to-school day in the West Village was a semiformal event. My outfit safely served as a protective shield from photographers' attention, which is how I found myself standing on the outskirts of the impromptu fashion show next to a little girl with barrettes stuck in her hair like pieces of gum who was leaning against the side of the school wall like a nonverbal Rizzo. Suddenly her mom broke from the photo-taking frenzy and beckoned me with a curled finger.

"Could you watch her?" Fashion Mom said to me, pointing at Barrettes Child and eliciting a command more than asking a question. Clearly someone wanted to make the most of their back-to-school Insta-runway moment and didn't need the distraction of their own progeny to ruin it.

"Sure," I said. Why not? What did I care? I could watch her kid. Barrettes Girl crossed one leg over the other like a sheriff. She looked me up and down in the same manner that her mother had and then she immediately took off for the school's entrance.

"No, you need to watch her!" Fashion Mom turned around to shout at me. "You need to watch her or she just runs off!!!" At that, Fashion Mom broke her conversation with an actual movie star and bore the burden of retrieving Barrettes Girl herself. "God," she said huffily to me, "I hope you're more aware with your employers' kids."

Oh! I realized. She thinks I'm the *babysitter*. I tried to hide my amazement as I processed the information, and thankfully I had the good sense to try to embarrass her back. "Who are 'kids'?" I asked Fashion Mom very loudly, trying to sound both inept and foreign. I smiled and made my eyes two different sizes, a talent caused by a slight birth defect that has proven to be immensely useful at specific moments in New York. At first I was pleased no one would ever ask me to watch their kid ever again. Later on I wondered if maybe the mom didn't know who I was specifically because I had largely opted out of the parent-on-Instagram rituals. Fashion Mom not knowing that I was a mom was probably my own fault. Maybe I should get on the mom social media train after all.

In the 1990s the Internet was my respite from the people in my physical life. America Online was all about weak ties, about talking with people you would never see in your life, which created a specific, often strange version of intimacy. Nowadays social media is about strong ties, which can make it distinctly opposed to being intimate. The boss's boss's boss is also a mom on Facebook. It may be called "friend-ing," but it's nothing like my mom by the brown town pond, because that stuff is reserved for real life. A few years in I realized I was simply lonely. I didn't have any mom friends. It was just that having so many moms friending me online made the loneliness a lot harder to spot.

A couple of years into parenting, our hallway neighbors invited us over for dinner. Brian and Elodie had two children: a one-year-old named Jasper, who was mischievous and beautiful, like a Jonathan Adler–designed doll, and a four-year-old named Eva, who was fully bilin-gual and disconcertingly good at decoding any adult conversation not intended for her consumption. From our sparse hallway interactions these people all seemed like fun. Cautious, I took note of their names and reverted to a Google search once back in the safety of our home. I sighed with relief when I found out their professions: photojournalists.

Photojournalists couldn't support the NRA! Perhaps we were finally on common ground.

The evening of our dinner we walked in to warm greetings and Jasper happily smearing something shiny and viscous all over the floor. In the corner Elodie was readying a mop. "What spilled?" I asked, making some light conversation.

"Meat juice," said Elodie, her heavy French accent making it sound not at all like something tiring but rather something fabulous.

"Meat juice?" Lauren was confused. It seemed that Jasper had gotten into the packages of ground beef that were resting on the counter, waiting to become tacos. There wasn't time for more questions, though, as Jasper was already being air-lifted into the bathtub. Remains of the meat juice sparkled in the candlelight.

Once Eva and Beckett realized that they were being upstaged by a younger child, they each invented their own personal crisis and competed heartily for the available adult attention in the room. Beckett emerged from Eva's bedroom having made his face up with Mr. Sketch markers; Eva read *The Little Mermaid* in French while taking frequent breaks to look at herself in the mirror and bow. Over the course of the next forty-five minutes Jasper sent a cart of finished tacos sailing into the wall and repeatedly punctured a leather couch with safety scissors, Eva announced that she'd lost her unicorn magic and might never get it back, which caused her to sob inconsolably for twenty minutes, and Beckett bounced a basketball off the flat screen television. Toward the end of the night Jasper danced merrily around his mother's legs as she tried to take a chocolate cake out of the oven. It goes without saying that the chocolate cake was quickly thrown on the floor.

Once the kids were medicated with a movie, I pulled an IKEA chair that had recently been scraped clean of glitter and frosting out to the deck to watch the sun set over the city. Elodie brought out a bottle of wine. Immediately we began to laugh about all the things that had happened, all the kids' various over-the-top moments. The night ended suddenly due to some crisis or another, and we dragged our kids to bed. *Time to go home.* The whole thing happened so quickly that it took me until the next day to realize that I had just found my first parent friends.

I now have our own little social network going with Elodie, Brian, and Lauren; we ping back and forth over texts, emails, Facebook posts, Instagrams, and other ephemera. We have truly shared our lives in the most New York of senses, meaning that we left our adjacent apartment doors constantly open so our kids could run in and out, trashing each house in succession. And man, do we manufacture Instagram sausage— it turns out that even photojournalists can't resist the fun—and those images have proven to serve their place. They become hallway posters, coffee table books for grandparents, birthday mugs, or just a quick way to turn back the clock when stuck in a taxi, waiting for traffic to clear. I'm not as averse to the whole parent social media thing as I once was. That said, as for making socially ambitious online connections with all those other tony moms in my city, I'm never going to be that person.

My life is largely a byproduct of being a lucky member of that first generation of people online, as rudimentary chat rooms and comically designed websites did a massive amount to help me feel less alone, to help me find people like me. Random online encounters eventually determined where I went to college, what type of employment I pursued, even who I married. It was quite a surprise at twenty-nine when I became a stepmom (and a gay stepmom at that) because it was the first time in my life when there was *nowhere* to go on the Internet for any sort of specific community—absolutely nothing. *Stepmoms? Gay ones? None here!* the Internet shrugs. *Just blend in with the other moms!*

The Internet is normally so good at fostering communities with very singular, specific problems or interests, it's come at the detriment of forming broader communities. We don't have great ways to disagree with each other or understand our differences online. Our biggest social networks, largely unchanged for the past decade, have gotten long in the tooth. Perhaps venture capitalists have been hosting too many barbecues, as they have proven unable to fund the types of invention that our world really needs. Or perhaps it's on us to ignore the venture capitalists and remember that the Internet doesn't create our values—it reflects them.

It's time for the Internet to evolve again. And it will. There is someone like my dad at a school right now, demoing a new communications tool to an audience of onlookers who only partially understands the glory of its implications. And what a memory that will be when that audience reflects on that moment twenty years from now.

Acknowledgments

Thank you dear, patient original family: Dave, Rob, Annie, and Kris Kimball. Who would have thought that Lego Town ended up a part of the paperback cosmos? Absolutely no one. And my new family: Papa Hudak, Megan, Gary, Jen, Dave, Papa, Erin, James, Lauren, and Beckett—I love you to the moon and back.

Jumana Abu-Ghazaleh, your whole world is running on the fumes of your encouragement—and they are potent enough to yield Oscars. Thank you for being my book-writing cheerleader ten years ago.

Marc Shaiman and Scott Wittman, thanks for lifting me from trash to paradise.

MB/JWB/SJ, there will never, ever be words for how much I love you people. Call me if you ever need to get out of jail.

Lauren Sharp and the Kuhn Projects team, you were my "reach school." Lauren, you are everything a writer could want in an agent.

Jess Fromm, you are so positive and wonderful as an editor—I still marvel at how you never told me any of my terrible ideas were terrible. And yet they still went away. You made every day of working on this book such a delight.

Josephine Mariea, your editing work is art. Michael Clark, thank you for keeping this project on track.

Greg Kimball, Margaret Kimball, Lucian Piane, Lis and Kilian, Sylvie Rosokoff, John Fischer, John Hill, Ben Jenkins, Audra Gullo, Sam Herschkowitz, Danika Isdahl, Ronit Wagman, Clive Thompson,

Steve Eisman, Carla Hendra, Jia Tolentino, Rebecca Milzoff, Hamilton Nolan, Angeline Koo, Elodie Mailliet Storm, Brian Storm, Aaron Kennedy, Anya Stiglitz, Lecia Sequist—what a list of big shots! I must have done something significantly impressive in a past life to have people like you in this one.

And lastly, one more time, thank you, Lauren Leslie, for putting up with my life-threatening dishwashing allergy, my hot sauce addiction, my junk bowls, my junk drawers, and the mentally unstable dog I bought from a website. You are my angel.

About the Author

Jess Kimball Leslie lives in New York City with her wife and son.

Photo Credits

Device photos by Sylvie Rosokoff.

Wint (@dril) tweets printed with permission.

Myst screen: This image is copyrighted by Cyan Worlds Inc. Myst ™ is the sole property of Cyan Worlds, Inc. Copyright 2003 Cyan Worlds Inc. All Rights reserved. Used with Permission.

CPSIA information can be obtained
at www.ICGtesting.com
Printed in the USA
LVOW13s1925231216
518612LV00001B/1/P